T0213619

SpringerBriefs in Applied Sciences and Technology

Computational Intelligence

Series Editor

Janusz Kacprzyk, Systems Research Institute, Polish Academy of Sciences, Warsaw, Poland

SpringerBriefs in Computational Intelligence are a series of slim high-quality publications encompassing the entire spectrum of Computational Intelligence. Featuring compact volumes of 50 to 125 pages (approximately 20,000-45,000 words), Briefs are shorter than a conventional book but longer than a journal article. Thus Briefs serve as timely, concise tools for students, researchers, and professionals.

More information about this subseries at http://www.springer.com/series/10618

Alexander Raikov

Cognitive Semantics
of Artificial Intelligence:
A New Perspective

 Springer

Alexander Raikov
Russian Academy of Sciences
Institute of Control Sciences
Moscow, Russia

ISSN 2191-530X   ISSN 2191-5318   (electronic)
SpringerBriefs in Applied Sciences and Technology
ISSN 2625-3704   ISSN 2625-3712   (electronic)
SpringerBriefs in Computational Intelligence
ISBN 978-981-33-6749-4   ISBN 978-981-33-6750-0   (eBook)
https://doi.org/10.1007/978-981-33-6750-0

This Springer imprint is published by the registered company Springer Nature Singapore Pte Ltd.
The registered company address is: 152 Beach Road, #21-01/04 Gateway East, Singapore 189721, Singapore

Dedicated to my parents and teachers

Acknowledgments

The author is grateful to his wife for her incredible patience. Also, the author is grateful to Alexander Ageev, Vyacheslav Abrosimov, Zinaida Avdeeva, Yuri Bundin, Yuri Hohlov, Vladimir Lepskiy, Dan Lubarsky, Alexander Mazurov, Dmitry Novikov, Alexander Ryjov, Massimiliano Pirani, Viktor Saraev, Sergey Silvestrov, Mikhail Sazhin, and Sergey Ulyanov for their participation in discussions of the results and support.

Introduction

For 2 million years of formation, humankind has already understood itself well enough for creating a similar kind—artificial intelligence (AI). At the same time, for over 100 million years of their development, the ants have acquired skills the nature of which humans can only guess at. Moving towards a goal, a human leaves knowledge on his or her way, whereas an ordinary ant leaves a chemical trace. Human knowledge is accumulated and lost, whereas ant's traces first increase and then evaporate.

The disappearance of knowledge and traces forms feedback needed for all living things on the Earth to govern, survive, and develop. However, there are some kinds of ants leaving no traces, but they orient themselves with respect to the terrain quite well. The navigational abilities of such ants are not yet very clear to humans. Perhaps they compose in their brain a cognitive map of the terrain, reflecting various marks, or maybe they are guided by the magnetic field of the Earth or by cosmic luminaries.

The study of human knowledge and consciousness naturally comes across even more complex questions. For example, a human has unique characteristics: the soul, the capability to think and intuition, emotions and feelings, and the conscious and unconscious states. A human tries to represent consciousness and thinking in the form of a cognitive system. Its behavior is determined by the dynamics of many objects interacting with each other, which are lost and found. Apparently, the behavior of a human thought can also be described in this way. However, such a discrete representation of thinking does not provide a complete understanding of the phenomenon of consciousness, subjective experience.

A human sees AI as such a cognitive system. Despite a relatively long history of this term, AI has not yet encroached on many of the cognitive abilities of humans. The reason is that modern AI does not go beyond formalized systems and discrete logic, even the most complex one entangled in deep artificial neural networks. AI is immersed in an electronic computer and an iron robot, both having neither human soul nor ability to suffer and empathize.

Many publications, including the ones of philosophical and historical character, have considered the relation of the human soul and subjective experiences to the algorithms of behavior. The result of such considerations is that an algorithm cannot replace the soul. If one intuitively admits a scientific representation of the soul, then attempts to describe it in logical terms will immediately turn the soul into an inanimate scheme. Conversely, the incorporation of the soul into logic denies the latter. In a computer, logic cannot be used to interpret the "Mental experience" operator deeply enough.

At the same time, the issue of duality and integration of the purely human and computer attributes is permanently relevant. After all, one day or other, the progress associated with increasing the density of the number of transistors in a computer processor will reach the atomic limit. Now the size of a transistor has already decreased to 10^{-8} m, while atomic dimensions start from about 10^{-10} m. Note that the dimensions of an atom are much less than the wavelength of visible light (nearly 10^{-7} m). Therefore, they cannot yet be seen with an optical microscope, unlike transistors. Attempts to see an atom through such a microscope encounter fundamental limitations due to an increased distortion of information about the atom with an adequate reduction in the wavelength. In addition, it should be emphasized that a transistor is much simpler than a human neuron. The former has three terminals, while the latter can span one axon and thousands of dendrites for communication with other neurons. What is of fundamental importance, a transistor responds to the same input signal in the same way, whereas a human neuron responds spontaneously.

Human consciousness and thinking cannot be completely described in the language of transistors. Therefore, when creating AI systems, the purely human characteristics are taken into account by developing hybrid technologies that involve both a human and a computer: a human enters a task into a computer and performs corrective iterations (if necessary) at different stages of solving the problem. These iterations can be quite complex, e.g., in accordance with inverse problem-solving methods. In such cases, small variations in the initial data may cause considerable changes for the resulting solution.

Thus, two types of semantics can be associated with the symbolic models of AI, i.e., two types of mappings of these models into external objects and events. The first type is represented by formal systems, neural networks, logical constructions, discrete images, etc. All these form the so-called denotative semantics. The overwhelming majority of modern AI methods use this type of semantics. The second type of semantics cannot be represented by any formalized scheme and logic. Such semantics is called cognitive. Cognitive semantics is nonlocal in some sense, lying outside the symbolic model, and its formalized extensions.

Due to two types of semantics, there are two different ways of human decision-making. On the one hand, a human can formulate decisions in a logical fashion; on the other, he or she has the ability to evaluate things using emotions. One is deliberative and logical; the other is affective and driven by the feeling. Both systems interact to control human behavior. The first system is associated mainly with the construction of denotative semantics and the second with cognitive semantics.

This book mainly considers cognitive semantics that is not directly formalizable, attempting to represent them in an indirect way. For this purpose, inverse problem-solving methods in topological spaces, ideas of controlled chaos, wave theory, quantum field theory, the theory of relativity, some elements of astrophysics, mathematical methods of category theory, and group theory are used.

The theoretical aspects of constructing cognitive semantics in this book are explored with application to problems of practical interest. For example, it is required to accelerate the development of a corporate strategy dramatically or to conduct a group brainstorming session under critical conditions. The strategy in a corporation is usually elaborated in several months, at best, in several weeks. However, an urgent task can be to construct an action plan in a few hours, or even in a few minutes, when facing emergency situations. In this case, AI methods should be used in order to create the appropriate conditions for accelerating the strategic process and making it purposeful. A typical brainstorming session takes 2–3 h. As a rule, it is divergent in nature and intended to generate useful ideas. A new formulation of the task may require simultaneous multiple brainstorming sessions on a specific issue. Moreover, the process should converge.

Problems that are even more difficult include the explanation of phenomena such as quantum gravity, dark energy and dark matter, and proof of cosmic strings. Apparently, these problems should be solved in a simultaneous and coordinated way by several groups of researchers from different areas of science and different countries. AI can help to integrate and synchronize such collaborative research. However, modern AI tools, such as deep neural networks, knowledge management, and ontological models, will clearly not be enough. After all, modern AI models cannot simultaneously take into account the entire interdisciplinary context of the problems under study: AI is not immersed in the world, being often unable to explain the results of its actions.

The cognitive semantics of AI is interdisciplinary. This book attempts to show that while maintaining the inconsistency and uncertainty of the phenomenon of human consciousness, cognitive semantics can be employed to cover, in a much more holistic way, various aspects of the problems and tasks solved using AI, both in the macro- and micro-worlds, in the cosmic and quantum scales.

Contents

About the Author

Dr. Alexander Raikov is Leading Researcher of the Institute of Control Sciences of Russian Academy of Sciences, Doctor of technical sciences, State Advisor of the Russian Federation of the 3rd class, Winner of the Russian government award in the field of science and technology, Professor of the MIREA – Russian Technological University, and Head of the artificial intelligence department of the National Center for Digital Economy of the Lomonosov Moscow State University. In 1992–1999, he is Chief of the analytical technology department of Russian President Administration. His scientific and business interests are in AI, decision support systems, and strategic management. He published 7 books and 450 papers in periodical journals and proceedings of conferences. He focuses on the advantages and risks of strong AI.

Abbreviations

AC	Atomic component
AGI	Artificial General Intelligence
AI	Artificial Intelligence
CMB	Cosmic Microwave Background
CS	Cosmic string
DNA	Deoxyribonucleic acid
EB	Electronic Brainstorming
EBK	Einstein–Brillouin–Keller
EPR	Einstein–Podolsky–Rosen
GUT	Grand Unified Theory
H	Hadamard matrix
HBT	Henbury–Brown–Twiss
HO	Hadamard operator
LHC	Large Hadron Collider
MHF	Modified Haar Function
OC	Optical computer
QC	Quantum computer
QGA	Quantum genetic algorithm
QS	Quantum semantics
UFT	Unified field theory

Chapter 1
History of Knowledge

The history of knowledge begins with the moment about which humankind has actually no knowledge. The phenomenon of knowledge is usually associated with human history, but it can be even more ancient. Its formation dates back to the birth of the Universe. In this case, knowledge is a much more complex phenomenon than the one adopted in modern Artificial Intelligence (AI).

For knowledge representation, traditional AI uses symbolic forms, including natural language, logic, and neural networks. In fact, knowledge can be represented taking into account the relic processes of the formation of matter and energy, including their components at the atomic level. In this chapter, let us take a look at the brief history of symbolic forms of knowledge representation to show in the following chapters that these forms are very restricted.

As is believed, *Homo sapiens* emerged around 280–300 thousand years ago. But only 50–12 thousand years ago (the Upper Paleolithic age), humans began to diversify knowledge. This could be due to various reasons, e.g., the need to accumulate skills and knowledge for catching birds [1]. Humans started creating knowledge in the form of artefacts, campsites, cave paintings, petroglyphs, carvings, and engravings on bone or ivory.

Then the Agricultural Revolution shifted human lifestyle and cultures from hunting and gathering to agriculture and settlement. As a result, humans experimented with crops, learned, and accumulated knowledge about cultivation.

Since ancient times, legends and folk books glorified knowledge. The heroes of legends, performing great feats, selflessly obtained knowledge about the sources of life. As legends say, new knowledge emerges from magical dreams.

At that time, fidelity, love, and knowledge were of value: for obtaining something, a riddle was made, and a task was given, as physical strength was not enough to succeed. And everything went well thanks to knowledge. The spells were important for survival in those ancient times.

Slavic amulet Garuda opened the secret knowledge to worthy people in order to acquire the gift of prediction [2]. Different prescriptions were captured in aphoristic

© The Author(s), under exclusive license to Springer Nature Singapore Pte Ltd. 2021
A. Raikov, *Cognitive Semantics of Artificial Intelligence: A New Perspective*,
SpringerBriefs in Computational Intelligence,
https://doi.org/10.1007/978-981-33-6750-0_1

cuneiform on clay tablets. This is the first knowledge preserved with a newly invented writing system in the Ancient East.

The Ancient Middle East distinguished and structured Vedic knowledge [3]. There are four levels and six states of consciousness. The levels of consciousness are as follows: Words, Thoughts, Feelings, and Meditations. The states of consciousness are as follows: Sleep, Dream, Wake, and the Transcendental state (Contemplation, as the intersection of Sleep, Dream, and Wake), the Conceptual state, and the Cosmic state.

Already in those distant ages, there were several pillars of knowledge. The pillars helped humans survive in difficult conditions. One system of four pillars of knowledge will be described below. This knowledge contains much that cannot be expressed in words: it is necessary to think out the details, taking into account the current situation.

The first pillar of knowledge is discrimination. It is the ability to differentiate between changes. Everything in the world is changing. Today, it may happen that a human experiences a feeling of loss, resentment, or pain. However, everything changes, and tomorrow these feelings will disappear. A human is unhappy every time for a different reason. The reasons are changing. When a human realizes that everything is changing (winter replaces autumn, bad becomes good, etc.), he or she becomes stronger through adversity. Knowledge is accumulating. Knowledge is very simple; it is the simplest and most accessible thing. However, this means that knowledge is difficult as well: potentially available, but not available in dynamics. This is how consciousness is arranged: first of all, it is able to see and distinguish changes.

The second pillar is dispassion. If a human is in misfortune, he or she begins to live with hope. The unhappy live with hope. The unhappy try to believe that everything will be good. Nobody can live without it. However, hope can grow into passion, and passion can completely capture a human, making him or her a sliver floating in a turbulent sea of feelings, thrown in different directions by the waves of emotions. Dispassion is neither the absence of emotion nor apathy. It is not depression. Quite the opposite, dispassion is the absence of emotional fuss; it is enthusiasm. Events and situations are constantly changing, and these changes can happen very quickly. Consciousness smoothes out these changes, not allowing a human to be tied to the chariot of effects. Consciousness prevents from doing rash deeds. Consciousness is dispassionate. It perceives events without feverishness, and at the same time, with concentration and seriousness.

The third pillar is the six kinds of wealth. The first wealth is the ability to concentrate consciousness. In crucial moments, consciousness focuses on the main thing. The second wealth is the ability of consciousness to rule over feelings. With this ability, a human will not become obsessed with distracting desires, thereby avoiding stress. The third wealth is perseverance or patience: with this ability, human consciousness remains stable when an opposite event affects a human. The fourth wealth is the ability to enjoy one's nature. It promotes the natural manifestation of real feelings, rather than toadyism and servility. The fifth wealth is faith. Faith and trust, without which life is impossible: an unexpected event may always occur. The sixth wealth is serenity. Consciousness knows how to recreate, easing the body. This gives rest, relaxation, and the release of unused forces. The six kinds of wealth are the potential serving to move a human along the right path.

The fourth pillar is the desire for freedom and enlightenment, the inevitable intentions of a human, as long as he or she can remember him/herself. This is how the sages of the great Vedic civilization saw the foundations of knowledge 4–5 thousand years ago.

For humankind, the emergence of writing in the late 4th–early first centuries B.C. is comparable by importance with the conquest of fire. Fire left behind the time of savagery, and writing gave humans the power of knowledge to follow the road of civilization. As was written at that times, "Wise actions start from knowledge, not from guess" (Aeschylus [4]). Condemning recklessness as the cause of troubles, evils, and misfortunes, declaring the indisputability of the idea of justice, seeing arrogance and contemptuous pride as the cause of violated justice, Aeschylus concluded that mind, thought, and knowledge are the greatest good. The goal was to get knowledge for people, to enlighten them through the gift of mind and ingenuity. "Wisdom is the oneness of mind that guides and permeates all things." (Heraclitus [5]).

At the same time, the tragedy of some lies in ignorance (the lack of knowledge), and the tragedy of others in perception (the excess of knowledge). And already in those days, it was clear that the tragedy of perception is perhaps the most difficult and terrible since it plunges a human into an abyss of contradictions and spiritual torment: "Knowledge, knowledge! A heavy burden when, to the detriment of those who know, you are given!" (Sophocles). The deed of Prometheus freed people from ignorance and brought the wrath of Zeus on him. With knowledge, part of the divine power from the gods is transmitted to people. However, the victory of will and mind in favor of justice is higher than a belief in divine law, causing no conflict.

What knowledge did the founder of the world religion of Buddhism bring to reduce the suffering of people? The central element of knowledge in this religion is the four noble truths associated with the four elements of human existence: suffering, the cause of suffering, the end of suffering, and the path that leads to the end of suffering. They remain the same right now. What stimulates a human to make a decision? It lies deep inside him/her. Nevertheless, in outward appearance, each decision has four components: discomfort, its cause, the goal to eliminate it, and the path to achieving this.

Two thousand years ago, a logically accurate definition of rational in thinking began to emerge. This was associated with the name of Aristotle. He was the first to discriminate between the intuitive and the analytical in thinking. From the thoughts of his predecessors, Aristotle extracted, first of all, the sign of the changeability of nature. In changes, he separated matter from the form as well as abstracted the form from its means of representation. With this approach, now it is possible to simulate and perform calculations on computers over the form (numbers, symbols, and formulas), not forgetting about its correlation with reality, with emotions. Aristotle laid the foundations of cognition, which recently has become of consumer value as methods of system dynamics and cognitive modeling. Aristotle believed that knowledge is based on the study of thought.

This idea of Aristotle gave birth to the modern science of the meanings and truth of sign systems, called semiotics and hermeneutics. Recall the Aristotelian discourse. There was no concept of an event in it. However, the modern interpretation of "event" matches the following triad considered by Aristotle:

Potentiality—Action—Actuality.

This triad constitutes a whole, i.e., a closed and self-contained wholeness. The domination of Potentiality allows for fantasy. This is a hypothetical virtual space in which nothing is prohibited, for example (today):

- the energy of a cosmic string is by 12–15 orders of magnitude higher than that imparted to a quantum particle by the Large Hadron Collider (LHC);
- a car-sized annihilation engine that can energize our entire planet;
- a nooscope that sees what is happening deep in the consciousness of humans and the Universe, thereby being able to generate brilliant thoughts, etc.

With the domination of Actuality, there is the reliance on common sense, laws, and regularities, and behavior is subordinated to the set of essential principles. However, only within this set, causes and stable forms are effective, and goals are logically constructed. At the same time, logic diminishes the completeness and depth of the presentation of a phenomenon, being split into analytical components. This means that in Actuality alone, goal-setting loses its integrity.

In the paradigm of the Aristotelian triad, the success of decisions considerably depends on the position of the middle component, Action. If it is located closer to Actuality, then each action is realized. In addition, Action obeys known or, for the time being, unknown laws and regularities and is controlled by logic. Mathematical probability works within the same logical stereotypes; it is constructed by choosing certain alternatives, known or axiomatically deducible.

The truth of the completely new lies beyond the border of knowledge. In any model, the main factor may not be visible but latent. Such a factor is fundamentally incognizable, indefinable by logic and probability. New is when Action moves away from Actuality, shifting towards Potentiality. Then the discourse of free will, intention, and unconscious dominates. Actuality becomes external to the source of voluntary fantasies. In this position, close to Potentiality, Action dominates and gives rise to completely new events.

In many systemic-epistemological constructions, there is a mysterious and logically incognizable Potentiality. For example, according to Hegel, Potentiality is a Concept; in Christianity, the Father and the Holy Spirit; in the Vedas, spirituality; in Kaballah, the Supreme World. And even in physics, Potentiality is hidden in the depths of the atom or in the mysteriousness of cosmic quasars. Potentiality as a whole cannot be seen or described with logic and words. As soon as it becomes visible (through a microscope), it is no longer Potentiality, but Action or Actuality.

Later on, the history of knowledge passed through the Renaissance, Machiavelli, Copernicus, Galileo, Descartes, Newton, Leibniz, and others. Skipping that time, let us address more recent lessons on the philosophy of knowledge.

References

1. Finlayson, C.: The Smart Neanderthal. Cave Art, and the Cognitive Revolution. Oxford. University Press, Bird catching (2019)
2. Garuda in Slavic Mythology and Buddhism. Great Power of the All-Devouring Sun - Mythical Bird Garuda Riding Bird Vishnu. (2020) https://iia-rf.ru/en/kids/garuda-v-slavyanskoi-mifolo gii-i-buddizme-velikaya-sila/. Accessed December 7, 2020
3. Shankar, R.: The Four Pillars of Knowledge. https://www.artofliving.org/wisdom/the-four-pil lars-of-knowledge (2016). Accessed December 7, 2020
4. Aeschylus: Agamemnon. https://www.poetryintranslation.com/theodoridisgagamemnon.php. Accessed December 7, 2020
5. Heraclitus: Quotes. https://medium.com/@nazmitarim/heraclitus-quotes-5610b68d5e6d Accessed December 7, 2020

Chapter 2
Philosophy of Knowledge

Aristotle, Descartes, Kant, Hege, Ampere, Heidegger, Husserll–this is a shortlist of philosophers who contributed their bit to the comprehension of the phenomenon of knowledge. For example, according to Hegel, consciousness is the spirit, as concrete knowledge; moreover, it is immersed in the external. Hegel defined knowledge in the context of several notions as follows: consciousness, spirit, objective and subjective logic, formal, thinking, purpose, being, essence, concept, and idea [1]. In his works, Hegel emphasized that the formalization of thought is permissible, but it cannot be unconditionally extended to thinking: the boundless formalization of thought will, sooner or later, lead to the idea of the disappearance of thought. Formal thinking lies down that contradiction cannot be thought of. Refracting the processes of development and cognition through the method of self-movement of concepts and self-transformation of the spirit, Hegel substantiated the limitations of the rational and empirical in the cognitive process. The former transforms the spirit into the dead entity, whereas the latter mortifies the spirit by splitting into many independent forces.

Hegel believed that essence grew out of being, whereas concept out of essence; reality is a material whole that exists by itself beyond thinking as a ready-made world. When considered separately, thinking is empty: it adjoins matter as a form from the outside, is directed by matter, and acquires some content only in matter. Thus, in the structure of thinking, reality claims to have some autonomy, which in principle cannot be given an adequate logical expression. An attempt to replace this autonomy with a formal model, on the one hand, creates the illusion that thought processes can be described in a visual and constructive manner and, on the other hand, makes a real thought process epistemologically degenerate.

Hegel's logical attitudes can be represented as an aggregate of three elements, namely, being, essence, and subjective logic (or concept); alternatively, thesis, antithesis, and synthesis. This decomposition can still be useful today for problem-solving using AI tools. Being has autonomy and is independent of the language used to describe or represent it; in addition, being is independent of the thinking of humans

© The Author(s), under exclusive license to Springer Nature Singapore Pte Ltd. 2021
A. Raikov, *Cognitive Semantics of Artificial Intelligence: A New Perspective*,
SpringerBriefs in Computational Intelligence,
https://doi.org/10.1007/978-981-33-6750-0_2

who want to make a change in the area of being. The being under consideration can be a country, a ministry, an enterprise, a human, a physical body, etc.

Essence lies between being and concept: this is the area of formal, linguistic, and imaginative structures. As a rule, the question of the truth of the constructed essence concerns its adequacy to the concept and being. Essence is a specific set of qualities mediated by definitions and represented by a certain communicative form. It can be a model, some text, an expression, or a graphic image in a sign system.

The transition from being to essence is the process of forming essence. This is modeling. In this transition, an essential image of being is formed. Further, essence is included in the environment of concepts, which does not necessarily happen in a rational way.

In Hegel's point of view, concept is the third denial of being and second denial of essence: concept grew out of essence. Concept is the world of human understanding and thinking. In the area of concepts, goals are formed, followed by ideas for their implementation. Ideas differ from goals in their greater tendency to provide a representation of goals in the area of essence. According to Hegel, idea is the identity of contemplation and concept. After all, the formation of goals is followed by their implementation. The result of achieving goals is new essence. While possessing a certain degree of autonomy, essence is not conservative. It unfolds and develops, which is represented by the transformation processes of essential structures and is realized as a concrete impact on the being of reality via some means.

In 1834 Ampere gave a classification of sciences, mentioning cybernetics the third as the science of current politics and practical management of the state (society). The word "cybernetics" originated 2000 years earlier from the Greek $\chi o\beta\varepsilon\rho\nu\omega$ ("goberno," government), denoting at that times an administrative unit inhabited by people. But for the Greeks, the word meant something more. "Goberno" is an object of management that necessarily contains people with their thoughts and bias. A ship without a crew is not "goberno." "Goberno" is a ship with a crew, and the captain who controls the ship is "gobernet." When applying this understanding to the present day, a robot considered separately is not "goberno"; however, a robot interacting with a human in a purposeful manner becomes "goberno." Thus, even in Ancient Greece, there was a science accumulating knowledge in the form of sets of management rules, which is now called "knowledge management."

Aristotelian logic did not immediately grow into a clear formal interpretation. This occurred later, thanks to the works of Frege, Russell, Gödel, Whitehead, Buhl, Escher, Babbage, Tarski, von Neumann, Hilbert, Heyting, etc. Modern computers are working based on the formal interpretation of knowledge they developed.

However, people are trying to use computers for solving non-logical problems, the ones that life puts before them. The logical correctness of a theory does not imply the existence of its subject: the latter can only be proved by experiment, at least, in physics. The truth of logic in real life can be intuitively justified, too.

In classical mathematics, contradiction is undesirable; in real life, it is natural. Meanings and movement originate from contradictions and paradoxes. Exploring logical paradoxes, Poincaré created the concept of constructing things as essences from other essences. His idea was that contradictions in logic arise from the use of a vicious circle of definitions: an object is defined using other objects the existence of which depends on the object under definition. He believed that this is unacceptable in mathematics. However, this is acceptable in physics, where the state of a quantum system depends on the observer.

Sharing the doubts of Kronecker, Poincaré, Gauss, Lebesgue, Borel, Weyl, and others regarding the correctness of the presentation of mathematical reasoning using traditional logic, Brouwer became the founder of mathematical intuitionism, a philosophical-mathematical trend that considers intuition to be the only source of mathematics and the main criterion for its rigor and constructiveness. This idea can now be used to create strong AI.

According to intuitionism, the subject of mathematics is mental constructions, considered as such, regardless of any questions about the nature of constructed objects, like the question whether these objects exist independently of our knowledge of them (Heyting). Mathematical statements in intuitionism are some information about the mental constructions performed.

Heidegger, and later Gadamer, believed that the world around us could not be rigidly delimited from a human since it is structured through his or her goals, circumstances, and intentions. This statement also appears to be crucial for the development of AI.

Following a phenomenological tradition expressed by its founder Husserl ("pure consciousness" is not mediated by subject matter), it is possible to establish a clear boundary between the ideal and the real, the immanent and the transcendent. At the same time, with such attitudes adopted in applied research, the connection between the individual components of cognition is lost: for example, the psyche is not taken into account. Husserl argued that accessibility to experience never means a simple logical possibility: it is always a human-motivated possibility within the whole. Only essentially related components can truly form a whole.

He introduced the method of phenomenological reduction: a naive life in experience with conducting theoretical research only should be replaced by performing real acts of cognitive propositions and ordering them according to the logic of experience, with further switching to a phenomenological attitude. This attitude suppresses the fulfillment of any cognitive propositions and "puts" the previously made propositions "in brackets." This attitude suggests performing acts of reflection aimed at the propositions made earlier, thereby comprehending them over and over again. According to Husserl, cognitive activity is carried out exclusively in the acts of the second (reflexive) stage, in the infinite field of absolute experiences as the main field of phenomenology.

In phenomenology, if a human really perceived something, it must be considered a phenomenon or an intuitive substance of something else, alien to him or her inside. If there is a reason for the phenomenon, it must be fundamentally perceived, not necessarily by a given human but by another one who sees and feels the situation better. Image or sign indicates something beyond it. Sign and image do not "express" object; they are alien to it. The physical thing is not alien to what is sensual-bodily, "expressing" itself in the latter (and a priori only in it). A sensually appearing thing

Fig. 2.1 Phenomenological interpretation of consciousness in context of AI

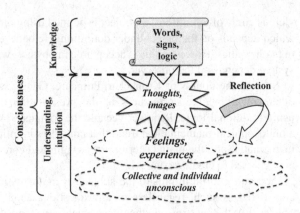

with its sensual form, color, smell, sound, and taste is by no means a sign of anything other than the sign of itself.

In Husserl's point of view, an experience-based way of comprehending situations prevails in the cognitive process. It performs the act of consciousness "attached to the body." Thus, bodily being is realized, and consciousness becomes an integral part of nature. In itself, consciousness remains what it is: its essence is absolute. This is due to the presence of *pure comprehending consciousness*. It realizes a double attitude: with one attitude, the grasping feeling reaches the object (thing), as if passing through its comprehension; with another attitude, it reflects this comprehension itself. In this case, as an object of comprehension, there can be a certain thought, feeling, image, or symbol, perhaps not even bound to a real physical thing.

A multilevel phenomenological interpretation of consciousness in the context of AI is shown in Fig. 2.1.

The phenomenological attitude, reflecting on and turning propositions off, can address pure consciousness, thus acquiring the form of an absolute emotional experience. Phenomenology does not deny reality; it only eliminates an unintelligible interpretation of the latter. The attitudes of phenomenology clearly suggest different semantics of AI models of controlled objects, namely, those that are alien to the signified, and those that merge with the signified and at the same time have a nonlocal character, sensually external with respect to the model.

Thus, over the years, the history and philosophy of knowledge have accumulated a methodological basis for the development of AI. First of all, this is an imperative that combines pure consciousness and emotional experience with rational and logic.

Reference

1. Hegel, G.W.F.: The Science of Logic. Transl. and ed. by George Di Giovanni, McGill University, Cambridge University Press, US, NY. 790 p. (2010)

Chapter 3
Nonlogic Thought

In traditional AI, knowledge and aspects of thinking are usually represented at the verbal-logical level: frames, ontologies, etc. At the same time, such phenomena as the causelessness of instantaneous insight and the illogicality of creative achievements [1, 2], demonstrate that logic is not enough to describe feelings, to express a thought completely, and to upload the collective unconscious into a computer.

The study of the phenomenon of reckless creativity leads to a reasonable idea of going beyond the logical-like representation stereotypes of cognitive processes that have been established in traditional AI. One of the approaches is to develop an ancient idea of conceptual thinking that involves an environment external to the human body. In this case, the bursts of creative insight allow for very unusual interpretations.

For example, the following model of consciousness and thinking of a human can be constructed. Parents pass to their child, along with genes, a uniquely organized set of subatomic particles, which predetermines human mental behavior. As is well known, the hereditary material contained in a cell of the human body is represented by 23 pairs of chromosomes. Human autosomes, sex chromosomes, and mitochondrial DNA all contain approximately 3.1 billion base pairs. These pairs consist of nucleotides on chains of nucleic acids linked by hydrogen bonds. The size of one nucleotide is 0.34 nm, being commensurate with an atom. At the same time, each atomic particle and its components are in interaction with the external environment due to physical fields and quantum non-local effects. Therefore, human consciousness and thinking can be treated as a closed system distributed in a fairly large space, including the cosmic and atomic levels. The neural component in this system, which is now the main focus in AI, plays an important but still particular role. Really, a certain supersystem of this closed system can be an observer of its behavior. In this case, human consciousness and thinking will be characterized by quantum effects, such as superposition, the collapse of a quantum state, nonlocality, etc.; see Chaps. 10 and 13.

This model can have various interpretations as follows: a thought is a particle of a certain universal consciousness; the brain is an antenna array with many resonant

© The Author(s), under exclusive license to Springer Nature Singapore Pte Ltd. 2021   11
A. Raikov, *Cognitive Semantics of Artificial Intelligence: A New Perspective*,
SpringerBriefs in Computational Intelligence,
https://doi.org/10.1007/978-981-33-6750-0_3

circuits that captures thoughts from the external environment; cognition is a physical field in all its known (and yet unknown) manifestations. The well-known fields include gravitational, electromagnetic, strong and weak ones; the little known fields are quantum gravity, dark energy, and dark matter.

If thinking is considered through the physical field hypothesis, then it must be described in the form of a complex infinite-dimensional space, with the reflection of radiation sources and receivers, the propagation of waves, and the changing states of quantum particles. In addition, there are relativistic effects contradicting the quantum ones. At the same time, as is well known, contradictions are permissible and even necessary in thinking.

The field nature of thinking, if accepted, suggests the idea to construct a unified field theory as a movement towards some ideal, which is convenient to represent a completely new reality of consciousness. Note that various interpretations of this process in terms of field theory, quantum theory, and relativistic theory, are possible.

Consider a particular case. A group of people is solving a problem: analyzing a situation, formulating a goal, constructing a model, thinking about it, making a choice, and elaborating a plan. At some time, a decision comes. However, such a time may occur too late: for example, a competitor will find a good solution faster. When people make decisions in a group, they tend to be bolder since group brainstorming activates the thinking of each participant. Meanwhile, the discussion of the issue is often divergent and can lead the collective thought far away.

A mysterious transcendental state of mind (the phenomenon of meditation), including individual or collective unconscious, helps to find a solution. This phenomenon cannot be represented in a metric space, and a concise definition seems to be almost impossible. These processes are non-formalizable. Their study requires an appropriate coverage of many scientific aspects of philosophy, psychology, physiology, physics, mathematics, field theory, the structure of elementary particles, and many others.

Meditation is closely related to behavior, discipline, practices, exercises, etc., bringing a human into a state between sleep, dream, and wake. The concept of meditation deals with such phenomena as relaxation (with the subsequent activation of mental activity), changes in metabolism, cure of diseases inaccessible to traditional methods, etc.

The unconscious often applies to the set of thought processes and phenomena that are not controlled by human consciousness. It is also studied by a wide range of disciplines: philosophy, psychology, psychoanalysis, psychiatry, psychophysiology, jurisprudence, art history, etc. For example, a psychoanalyst structures the mental activity of a patient in a special way for therapeutic purposes, using reframing techniques. The mystery of the unconscious inspires and realizes a human's hope for recovery from illnesses. A human expects from the unconscious some kind of impulse, generating a sudden and necessary intuitive decision or recovery.

When creating AI systems, the phenomenon of the unconscious in thinking is usually left out of consideration. At the same time, decision-making in a distributed environment requires the development of group modeling mechanisms based on physical and mathematical representations of this phenomenon. For example, in

Fig. 3.1 Strategy is solution of inverse problem

the paper [3], the game character of decision-making processes for the case of two competitors was studied. A method for solving the quantum Monty–Hall problem was used, which helps to anticipate the biased behavior of the engaged participant; also, see Chap. 10.

In the publication [4], an attempt was undertaken to speed up a strategic meeting based on an inverse problem-solving method on a topological space [5]; according to practical evidence, the attempt was a success. Such a statement of the problem arises due to its strategic character: the goals of the solution cannot be achieved by extrapolating the past management experience, and it is necessary to change the methodological paradigm of the solution (Fig. 3.1).

The inverse problem-solving process is incorrect: a solution may not exist; it may be non-unique, and, most importantly, may have no convergence to the goal. To solve such problems, the information generated by the participants during a meeting, together with their thoughts and feelings, must be structured in order to guarantee an accelerated convergence of the decision process to a consensus (collective agreed result); see Chap. 18.

The author's practice in moderation of strategic meetings in situational centers [4] shows a rather complex character of the interaction of participants when making collective decisions. This character is determined by interests (which can be understood) and the underlying psychological characteristics of interpersonal relations (which are difficult to identify). These characteristics arise in the process of coordinating positions and interests, especially when all group members demonstrate a desire to apply their efforts in the same strategically coordinated direction.

Each group member involved in a collective decision can have a different attitude, interest, and emotional state. He or she can hide them, or try to be open to other participants, showing tolerance towards the colleagues. When creating traditional decision support systems using AI, these aspects of participants' behavior are often assumed formalizable. In academic literature, they are even interpreted by formulas or simulation models with a linguistic, metric, or logical basis.

In fact, the hypothesis of a formalized representation of the decision-making process and the approach driven by it are far from the truth. The behavior of the participants is logically interpretable to minor part. Even in ancient times, different layers of consciousness were distinguished in the minds of people, for example, words, thoughts, feelings, and meditation; see Chap. 1. In addition, knowledge was usually of a communicative nature, structured in various ways, for example, in religious archetypal experience, myths, legends, and separation of the pillars of knowledge.

In collective decision-making, people participate with the following roles: leaders (top managers), hidden leaders, open opposition, employees, external experts, brawlers, and moderator. This list does not claim to be exhaustive. A special place is occupied by meetings, where three main roles are distinguished, namely, the group itself, leader, and moderator, who is responsible for reaching a collective consensus on the goals and ways of action, taking into account the leader's attitudes.

This decomposition of a decision-making group can be considered in the psychological paradigm: all group members are the collective patient, and the moderator is their psychotherapist. The group has a problem (being "sick with it"), and the moderator needs to help the group with "recovery," i.e., to solve this problem.

Thus, decision-making by a human or a group of people is, first of all, a volitional choice, a deep and purposeful release of emotions, transcendental insight. This process defies complete formalization. At the same time, researches into the development of tools to formalize the non-formalizable are carried out regularly; for example, see the attempts to formalize ethics in [6]. A most difficult challenge is to grasp the hidden dynamics of consciousness by means of formalization.

References

1. Perkins, D.: The Eureka Effect. The Art and Logic of Breakthrough Thinking. Norton, New York, London (2001)
2. Gigerenzer, G.: Gut Feelings. The Intelligence of the Unconscious. Viking, London (2007)
3. Raikov, A.N.: Holistic Discourse in the Network Cognitive Modeling. J. Math. Fundam. Sci. 3, 519–530 (2013)
4. Raikov, A.N.: Convergent Cognitype for Speeding-up the Strategic Conversation. In: IFAC Proceedings Volumes, vol. 41(2), pp. 8103–8108. Seoul, South Korea (2008). https://doi.org/10.3182/20080706-5-KR-1001.01368
5. Ivanov, V.K.: Incorrect Problems in Topological Spaces. Siberian Mathematical Journal, 10, pp. 785–791, (Novosibirsk, Russia, 1969)
6. Lefebvre, V.A.: Algebra of Conscience. Springer Netherlands, Series: Theory and Decision Library A, 30, vol. XIV, (2001). https://doi.org/10.1007/978-94-017-0691-9

Chapter 4
Hidden Consciousness

In the book [1], adhering to Jungian foundations, several features of human interaction without any classical formalized interpretation were identified. Such features include a hidden phenomenon called *the fusional complex*: a participant feels drawn to some object (subject) in a very painful way, and intimacy becomes neither a positive union nor the state of mutual dissolution. Moreover, magnetically sticky intimacy can exist under complete disconnection. This complex creates conditions inconsistent with rational thinking. A human has the feeling of connectedness (fusion) with someone or something in an intricate way; at the same time, there is no real connection. An attempt to eliminate such a fusion dramatically increases anxiety, fear, and the sense of fall. For example, this state can occur when a participant feels that the group no longer wants to work with him/her, but no one clearly shows this attitude.

This state of a participant (especially if hidden from other group members) cannot be displayed in spatiotemporal coordinates in the form of factors, concepts, and designated relations. As a result, the applicability of advanced information and analytical methods becomes considerably restricted. These group members seem to have positive experiences in earlier or parallel object relationships. And such experiences can contribute to a positive creative attitude of the participant. At the same time, incorrect actions of the group leader can pose an apparent threat for the participant to be separated from the group. Responding to this threat, the participant can address deeper levels of the fusional complex. In order to somehow mitigate his or her fate, a participant can perform an illusory decomposition of the object of fusion into good and bad parts, thereby contributing to the further disruption of his or her consciousness (as a consequence, to the further disruption of the consciousness of the leader and the group itself).

Interactions with such participants can greatly influence the leader, turning him or her into a state of confusion and making his or her position vulnerable to the success of the entire group. The leader faces with the difficult task of realizing his or her own tendency to decompose the whole, imposed on him or her by the participants. The

A. Raikov, *Cognitive Semantics of Artificial Intelligence: A New Perspective*,
SpringerBriefs in Computational Intelligence,
https://doi.org/10.1007/978-981-33-6750-0_4

multiplicity of parts of the whole, generated in this way by the corresponding participants, will contribute to the multiple decomposition of the leader's consciousness.
In this case, the leader's task is not just to follow a formalized plan of actions, but
also not to lose his or her identity, to reflect and experience individual participants
and the group as a whole.

This is a difficult task since the participants hide their feelings of fusion, do not
express their intentions. Moreover, sometimes they are even unable to understand
their feelings and intensions. A group member with such a fusional characteristic can
perceive, at a deep level, other group members or the leader so that this knowledge
turns out to be offensive, slighting, and mortifying his pride. For example, this can
happen when the leader becomes condescending or goes into a directive manner.

The universality and integrity of the problem solved within the group, apparently,
can be cognized only in parts and thanks to the joint efforts of the participants. Each
part generates a concept, its sign (symbol, name, factor), as well as semantic and
objective interpretations. Obviously, unity (integrity) exists, in one form or another;
otherwise, there would be no group. However, for a group facing a problem, the
aspect of integrity may be under threat, in the state of degradation, decomposition,
and dissipation. The leader's task is to assemble the parts into a whole and to engage
all participants in this process, at the same time giving the freedom of action and
creativity to each participant in his or her part. Thus, the space reflecting, to some
extent, the transcendental layers of consciousness is constructed with the following
set of features:

- the presence of hidden thoughts, intentions, parameters of the participants;
- the fusional effect, the attraction of consciousness to other objects at deep levels;
- the lack of similarity between the sign and phenomena of consciousness;
- the effect of deep intersubjective interaction;
- the limited character of the formalized discrete representation;
- the effect of explosive mental activity;
- the local and distant interactions of consciousness in categorizing phenomena;
- the capture of consciousness by the latent aspects in the behavior of participants;
- the need for integrity of the collective consciousness, etc.

Note that the features of consciousness listed above are important for the processes
of collective decision-making, but they cannot be given a formal representation. Any
attempt of formalization, even using a fuzzy or probabilistic measure, the substitution
of continuity by discreteness and irrational by countable, will distort the integrity of
both the phenomenon itself and its spatial representation.

In collective decision-making, there are proven approaches and methods by which
the group leader tries to ensure (a) the integrity of the phenomenon under study and (b)
reaching an agreement among all members of the group [2]. The sources of integrity
can be manifested from the depths of the transcendent, limitless, and collective
unconscious, as well as under the pressure of external influence dictating conditions
and needs. At the same time, the decomposition of the whole is provoked by the
manifestation of meditative activity and is stimulated by divergent intentions, when
participants freely generate their ideas and lobby their projects, thereby identifying

and showing their belonging to one or another part of the problem. Thus, group members tear the problem to pieces even more, on the one hand, plunging the leader into the state of torn consciousness, and on the other, giving rise to the chances of manifestation and development of new functions of integrity. Divergent processes are needed to generate collaborative ideas. However, these processes impede ensuring the integrity and convergence of collective creativity to an agreed result.

The meditative substance "encompasses" the problem being solved and the factors characterizing it as a flowing fluid, like the elements of any object or structure absorbing light. If this liquid-light analogy is considered as an approach to model the unconscious, then a wide range of issues should be taken into account, including such effects as turbulence, quantum shadow particles, decoherence, diffraction, interference, aberration, etc. If the liquid analogy is prevailing, then one should apparently address the topologies of hydrodynamics. As for the light analogy, quantum physics, the theory of relativity, and electromagnetic fields should be adopted.

Within the hydrodynamic analogy, the meditative substance will have to fill some bounded area of space and become incompressible. The space will be dynamic in nature, for example, a Hamiltonian space. This representation, in turn, can be used to ensure the stability of the decision-making process. As soon as the search procedure is completed, the final decision often acquires an ever clearer logical frame, forming a boundary between the external and the internal. In other words, the closedness of the resulting space can be supposed.

In this case, the meditative will be modeled by a set of smooth transformations of manifolds into themselves, like transformations in group theory. It can be the configuring space of an incompressible substance filling some area. This fluid-model framework allows using well-known mathematical tools to ensure a stable convergence of decision-making processes.

The human body, somas, neurons, and other elements interact with external electromagnetic fields. In particular, these interactions are used for research and medical purposes. Mathematically, this means that any movement of the external environment will influence a human like some morphism. Such an impact can change the inner energy of a human, his or her mental and emotional substances. Then, the space of solutions will be characterized by some topological structure, which implies preserving the volume of these substances under the impact of morphisms. This structure prevents the complete dissipation of the phenomenon, which is quite natural for the attributive existence of a thought or other source. On the other hand, the movement of an inhomogeneous medium of a body can stretch its elements, individual particles, and ensembles of particles, which will increase the internal potential energy of a human due to its transition from the kinetic form. This bi-directional process leads to a generator effect, creating electromagnetic fields induced by very small particles.

Speaking about small elements and particles, as an illustrative example, one can take the well-known hypothesis of shadow photons, which form the interference of De Broglie waves on the screen after the real photons pass two slits or a diffractive crystal. These shadow phenomena cannot be detected, but their assumed existence explains the effect of interference on a large ensemble of individual particles emitted by a laser with some periods of time. Then, with a degree of conventionality, the thought source can be represented as a laser: a laser beam performs diffraction on the "crystal" of the AI model, creating an interference pattern as the desired solution; also, see Chap. 10.

When developing this approach with an optical-laser analogy, the formation and processing of coherent optical radiation with repeated multilayer holographic signal recording may be of interest. At a fixed point of a holographic storage device not exceeding 1 mm in diameter, several hundred images can be recorded in theory, obtained by the interference of their Fourier transforms with a reference beam. Each image can be found relatively quickly through visual comparison with its small fragment and also reproduced on the screen by recording this reference beam. Operations can be carried out, literally, at the velocity of light (of course, data input–output operations being not taken into account.) Quite probably, the practical use of this approach is restricted by main factors as follows:

- There is the decoherence effect, i.e., the violation of coherence caused by the interaction of a quantum mechanical system with the external environment through an irreversible process, from the viewpoint of thermodynamics.
- An optical processor operates images, and images are, most likely, the elements of a thought phenomenon, not of meditative and emotional phenomena.
- The creation of materials for repeated multilayer holographic recording is a science-intensive engineering problem, unsolved yet.
- The processing of optical signals will be accompanied by large distortions (aberration, diffraction, etc.).

For the integration and cohesion of a decomposed whole, communication is required, which implies structuring and modeling, as well as a symbolic representation of the hidden. Apparently, any modeling or the use of signs represents the implicit, bringing out the attributes of what is hidden from full direct perception.

Classical AI tools can only work in a logical, formalized, and, most often, discrete environment. Even statistical and probabilistic methods are based, first of all, on the selection of a finite set of random events. Indeed, a course on probability theory often begins with a finite scheme considering finite probability spaces. The sets of events can form a Boolean or commutative algebra, be a simplex. Then the state of the system is described by a density matrix that satisfies certain relations. However, such a description, due to its matrix discreteness, is far from fully suitable for the representation of thought, emotional, and meditative phenomena. In this case, references to quantum analogies are inevitable.

The potential of quantum theory is aimed at displaying, studying, and predicting the behavior of the invisible and hidden world by means of classical representations. As it seems, the quantum–mechanical approach a priori cannot be considered sufficiently adequate to the solution of the problem under consideration: the physical reality of the world of particles, bosons, fermions, etc., can only partially be associated with meditative and unconscious substances.

The appeal to the quantum–mechanical basis touches upon a wide range of disciplines. For example, it is interesting to investigate the effect of a neutrino particle

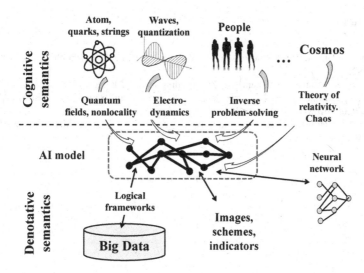

Fig. 4.1 Cognitive and denotative semantics

on meditative processes in consciousness. Neutrino belongs to the class of leptons; every second, about 6×10^{10} of them pass through an area of 1 cm^2.

In quantum theory, the problem of hidden parameters is usually solved relying on a basis of the Hilbert space, which is a generalization of the Euclidean space admitting of infinite dimension. Its inherent property is discreteness. The most important object in a Hilbert space is a linear operator. For any two elements in such a space, a scalar product and a complete space with respect to the metric generated by the scalar product are defined. A Hilbert space is a Banach space, i.e., a complete normalized vector space. Due to the discreteness and metrizability of a Hilbert space, it does not fully agree with the above-mentioned features of the phenomena of consciousness and unconscious.

In this context, the quantum-semantic interpretation of various levels in the space of consciousness can suggest how to improve the thorough consideration of the inaccessible while ensuring the integrity of a model representation of the problem being solved; for details, see [3]. Thus, for immersing AI models in the sphere of human consciousness, two main types of semantics should be separated out, namely, denotative semantics and cognitive semantics. The former is more associated with the material, logical, and formal representations; the latter is emotional and mental, i.e., unformalized. This separation is illustrated in Fig. 4.1.

According to Fig. 4.1, cognitive semantics is beyond the formalized capabilities of denotative semantics. For the cognitive semantics, other methods of interpretation have to be used. This book is devoted to the cognitive semantics of AI. For avoiding any temptation to describe the denotative phenomenon of thinking (i.e., the semantics represented using logic and formal languages only), it is necessary to make a small excursion into this area.

References

1. Schwartz-Salant, N.: Borderline Personality: Vision and Healing. Chiron Publications, Wilmette, IL (1989)
2. Raikov, A.N.: Convergent Cognitype for Speeding-up the Strategic Conversation. In: IFAC Proceedings Volumes, vol. 41(2), pp. 8103–8108. Seoul, South Korea (2008). https://doi.org/10.3182/20080706-5-KR-1001.01368
3. Raikov, A.N.: Holistic Discourse in the Network Cognitive Modeling. J. Math. Fundam. Sci. **3**, 519–530 (2013)

Chapter 5
Formalized Semantics

Consciousness and thinking produce objects of a formalized nature. They are usually discrete. In the depths of consciousness and thinking, the chaos of unformalized thoughts and emotions reigns. Their source lies at an even greater depth, which cannot be often described in a deterministic language. For example, consider a spontaneous change in the energy level of a quantum particle, which can be caused by the so-called zero-point vacuum oscillations (the fluctuations of a quantum system in the lowest-energy basic state). The supreme state of thinking is a natural language and formalized representations of the results of thinking.

A computer operates a discrete image, form, and logic. As has been discussed above, emotions have no form. A human can make AI guess something, formulate an epiphany or prophecy, or grow something like his or her intuition would do. In any case, AI will have to get this result and transmit it to the human. For this purpose, a language is needed. Language can be natural and formal. Sometimes, intermediate variants are introduced, e.g., the language of business prose, a limited natural language. People communicate in a natural language; computers and robots, in a formal language.

But neither evidence-theoretic reasoning nor model-theoretic reasoning considers the fact that the real process of reasoning, carried out by humans or machines, has strict internal limitations. The reasoning can be very diverse by nature, for example, non-monotonic reasoning, intended for the adequate operation of incomplete and changeable information. Fuzzy logic studies fuzzy reasoning; informal logic, informal reasoning. Hence, philosophical logic studies philosophical reasoning.

More than 50 years ago, category theory was created, one of the most important mathematical theories of the last century. This theory discovered completely unexpected relationships between various branches of mathematics and acted as a powerful tool for developing new foundations of mathematics.

One of the fundamental features of category theory is that it accepts "morphism" (mapping) as a primary concept on the same level as the concept of "object." In other words, according to this theory, there are only objects and morphisms between objects. Moreover, morphisms satisfy the laws of identity and associativity. All these together form a mathematical structure called "Category." With the elaboration of category theory, it became possible to create a logical universe much richer than "lattice," a structure whose elements are logics, in one sense or another.

A formal language includes the following notions. An *alphabet* Σ is a finite non-empty set. The elements of an alphabet are called *letters* (or *symbols*). A word over an alphabet Σ is a finite chain consisting of zero or more letters from Σ, and the same letter may enter a word several times. A chain consisting of zero letters is called an *empty word* and is denoted by λ. In a conventional interpretation, these notions refer to a dictionary, words, and sentences, respectively. The set of all non-empty words over an alphabet Σ is denoted by Σ^*. On this set, a binary operation satisfying the associativity law $(a \cdot b) \cdot c = a \cdot (b \cdot c)$ is defined (a semigroup in algebra). Also, this set contains an identity element (monoid).

If x and y are words over an alphabet Σ, then their concatenation xy is another word over Σ. This operation is associative, and the empty word is an identity element: $x\lambda = \lambda x = x$ for all x. Denote by $|x|$ the length of a word x. This function has the properties of logarithm: $|yx| = |x| + |y|$, $|x^i| = i|x|$. A morphism is a mapping $h: \Sigma \to \Sigma^*$.

There exist various classes of formal languages, e.g., *regular*. It is convenient to represent them in the form of *digraphs*. A digraph is a pair $G = (V, E)$, where V corresponds to a finite set of vertices and E to a set of ordered pairs of elements from V (*arrows*). The graphs are *labeled*: each arrow is assigned one or more letters from an alphabet Σ. Two subsets of V are also distinguished, called the *initial* and *final* vertices. Such a digraph is called a *finite state automaton* (or a *finite state machine*). A language L is called *representable* if there exists a finite state automaton G such that $L = L(G)$. A language $L(G)$ represented by a finite state automaton G over an alphabet Σ, consists of all words w over Σ such that w corresponds to some path between initial and final vertices.

Now consider the notion of grammars. A *substitution system* is a pair (Σ, P), where Σ is an alphabet and P is a finite set of ordered pairs of words over Σ. The elements (w, u) of the set P are called *substitution rules* (or *productions*) and are denoted by $w \to u$. *The derivability relation* $\Rightarrow$ on a set Σ^* is defined as follows. For any words x and y, the relation $x \Rightarrow y$ holds if and only if there exist words x_1, x_2, w, and u such that

$$x = x_1 w x_2, \quad y = x_1 u x_2 \tag{5.1}$$

and $w \to u$ is a production of the system. The reflexive transitive closure of the relation $\Rightarrow$ is denoted by $\Rightarrow^*$.

A *grammar* is a quadruple $G = (R, \Sigma_N, \Sigma_T, \Sigma_S)$, where $R = (\Sigma, P)$ is a substitution system, Σ_N and Σ_T are disjoint alphabets such that $\Sigma_N \cup \Sigma_T$, and Σ_S is a subset of Σ_N. Elements from the sets Σ_N, Σ_T, and Σ_S are called *nonterminal*, *terminal*, and *initial* symbols, respectively.

Each sign has a conceptual and object expression. This leads to the possibility of a dual representation for the relations of belonging. If an object belongs to the set of "Houses," then this object has certain properties inherent in the elements of this set, which make up a certain frame: the object has windows, doors, etc.

For example, the same term (word) can be in different relations to another term, so that the character of intensionality and extensionality is transferred to the relations. Let the symbol " $\in$ " mean "belongs as an element to"; also, let the symbol "ε" mean "has the property of."

The *kernel* of terms of a certain language is a notion expressing the property of a term formulated as "does not belong to itself as an element." *The kernel operator* maps a certain subset of language terms into a space in which all terms have this property: the convolution $\forall x (x\varepsilon F \equiv \sim x \in x)$ implies the formula:

$$F\varepsilon F \equiv \sim F \in F. \tag{5.2}$$

The right-hand side of (5.2) expresses the fact that the term F does not belong to itself as an element; the left-hand side, the fact that F belongs to itself in the sense of the property of not belonging to itself as an element.

Consider a formalism in which the formulas $x(y)$ ("y belongs to x") and $x = y$ are atomic, and the kernels of convolutions depend on one free variable and contain no parameters. Now let each atomic formula $x(y)$ be decomposed into atomic formulas of two types according to the scheme:

$$x(y) \diagup \begin{matrix} y\in x \ (y \text{ belongs to x extensionally}), \\ \\ y\varepsilon x \ (y \text{ belongs to x intensionally}). \end{matrix}$$

The formalism presented above is completely symmetric with respect to the symbols ε and $\in$. This symmetry is eliminated by introducing the axiom for the relations ε and $\in$, which can be written as

$$\forall x \forall y (y \in x \rightarrow y\varepsilon x). \tag{5.3}$$

In traditional terminology, it has the following meaning: if y is an element of class x, then y has a property that defines the class x. This formalism is asymmetrical with respect to relations.

Such schemes enrich a language by introducing notional parameters into it in a formalized way. In essence, a language is a "superficial" tool enabling people to communicate with each other, to transfer thoughts and emotions "in words." Note that the language of gestures and facial expressions goes deeper but remains "superficial" as well. Whatever the case may be, one can deepen to the level of emotional states of mind only using a language.

Constructivism found its most daring expression in the philosophy of *intuitionism*. This philosophy introduced a kind of intuitive audit (or filter) of mathematical transformations into the formalized toolkit. *A language is secondary* and serves for understanding in communication. Intuitionism admits formal systems as imperfect ones for describing knowledge. Such consideration of formal systems can be continued, for example, by addressing other modal logics. However, as emphasized at the very beginning, this book tries to escape a direct formalized description of thinking and emotions rather to treat them indirectly. One of such indirect approaches is the appeal to external spaces, including the cosmos.

Chapter 6
Cosmomicrophysical Thinking

When creating AI systems, thinking is usually associated with the human brain and the interweaving of its neurons. Less often, the matter concerns free will, instincts, and feelings, i.e., the connection of other human existence aspects and organs to thinking (the spinal cord, stomach, or heart). At the same time, the phenomenon of human insight is sometimes referred to as "gut feelings"; see the book [1]. The ancient thoughts about the cosmic and conceptual nature of thinking (the direct participation of objects from the external environment in thinking, including the distant ones) are recalled infrequently. Scientists usually take into account only the indirect impact exerted by external objects and phenomena on thinking through sensations, perception, and sense organs. The cosmic nature of thinking is more attributed to the area of astrology, in which, however, science is also trying its methods [2].

Against this background, the rare considerations on the nonlocal nature of thinking and its fractal nature, in a rather wide range of sizes (from the substantive level to cosmic scales), seem intriguing. Such considerations are well-grounded. The neurons of the human brain consist of atoms, and atoms of smaller particles. At the subatomic level, the behavior of particles depends on various fields, namely, gravitational, electromagnetic, strong, and weak ones. Gravitational field affects atomic particles, albeit very weakly. The gravitational attraction of two protons is 10^{37} times weaker than their electromagnetic repulsion. However, the number of atomic particles in the human brain is by tens of orders of magnitude higher compared to the number of neurons. Also, the expansion of vector and scalar fields into series includes infinitely many terms. Therefore, it is probably not worth discarding the small values of this field.

The perceptible effect of the Sun's prominences, the vagaries of the weather, etc., on the mood and thoughts of a human, suggests the idea to consider the impacts of both cosmic and quantum nature in the study of thought processes. The impact of the external space environment on the human brain and other organs can be treated as the receipt and transmission of antenna signals; see the discussion in Chap. 3. The human

A. Raikov, *Cognitive Semantics of Artificial Intelligence: A New Perspective*,
SpringerBriefs in Computational Intelligence,
https://doi.org/10.1007/978-981-33-6750-0_6

brain, body, and organs can be represented as a cognitive subatomic resonator. Each element in such a resonator (a cell, neuron, axon, dendrite, synapse, atom, quark, etc.) can be described by an oscillatory circuit with its own resonant frequency. This frequency can change under the influence of electromagnetic, mechanical, chemical, and other signals. As a result, various quantum-physical effects occur.

For example, it is possible to take into account the effect of entanglement: the particles located at very long distances are connected to one another; see Chap. 13. This effect cannot be explained within information theory since it contradicts the theory of relativity. To immerse the topic of thinking in cosmic and quantum physics, appropriate comprehensive research is needed, including investigations in the area of cosmomicrophysics [3]. As was noted in [4], Sakharov gave the idea that gravity is not "fundamental" in the sense of particle physics. He argued that gravity emerges from quantum field theory in roughly the same sense as hydrodynamics emerges from molecular physics. Likewise, the cognitive process can be considered both from the viewpoint of quantum physics and from the cosmological viewpoint, i.e., within the area of cosmomicrophysics.

In this area, cosmic and Earth-based interferometers and aperture synthesis systems have been created; the behavior of particles in the electromagnetic ranges is being explored; the anisotropy and spectra of relic radiation are being studied. In parallel, quantum optics is developing, which reveals the quantum features of optical signals received from the cosmos. Scientists are trying to discover the secrets of dark matter and dark energy. As a result, the structure of the Universe and the nature of gravity become clearer to humankind.

Computational experiments to establish a bridge between theory and observations are being carried out. These experiments are very complex and require energy resources that cannot yet be obtained on Earth using particle accelerators. The shortage of Earth-based energy resources for conducting physical experiments is about 10^{-12} compared to the capabilities of modern quantum accelerators. Quite obviously, experiments should be moved to the cosmos. For the time being, humankind is not ready for this, both due to high cost and an underinvestigated procedure of such experiments. For example, cosmic strings are a source of energy that cannot be achieved on Earth; however, even their existence has not yet been proven.

Constructing an adequate model of consciousness and thinking requires elaborating a unified theory of all fundamental interactions that will provide a complete picture of the interaction of particles, as well as explain the origin of the Universe and the appearance of radiation and matter in it. This model will cover at least a scale from the Planck length (10^{-33} cm) to the size of the Universe. Note that the apparent size of the Universe is believed to be about 13.75 billion light-years, and the real size to reach 45.7 billion light-years. Between the phenomena at different points of this global scale, the human body will occupy a kind of the "golden mean," whereas the space of his or her consciousness and thinking, apparently, will cover the entire scale. The microworld will be located to the left of the human's place, and the macroworld to the right of it. Conventionally, the former can be considered as the cause of thinking, and the latter as its consequence. There is a certain connection between the parts of this scale: atomic physics determines the properties of matter, biomass, and planets; nuclear physics, the properties of stars, black holes, and galaxies; the theory of relativity and physics of ultrahigh energies, the structure and behavior of the Universe. All these determine the substance, instrument, and motive of thinking.

Thinking and consciousness have tremendous power, which has yet to be understood. Nowadays, using AI tools, a text can be automatically edited to match the style of Shakespeare, a picture can be corrected to match the talent of Picasso. However, machines are not yet capable of writing poetry or painting a picture at the level of a human genius. Perhaps this will be achieved by mapping the harmony of the Universe into consciousness, which can be studied, in particular, within cosmomicrophysics.

The unity of weak, strong, electromagnetic, and gravitational interactions will fully manifest itself only at energies that are not reachable on Earth, for example, using ground-based particle accelerators. Such energies exist in the cosmos: this is the relic Universe. The imprint and energy potentials of this moment, for example, are carried by the cosmic strings [5]. Their study could give necessary information about the processes and time of inflation of the Universe, allowing to verify theoretical hypotheses regarding a unified field theory. As a consequence, this verification will help to approach the physical nature of thinking and consciousness.

The sources of data necessary to verify such theories are observations of the electromagnetic background radiation, its spectrum, and distribution in the celestial sphere, as well as the study of light from distant galaxies, the analysis of distorted light trajectories and light splitting on the way to telescopes, on the Earth and in cosmos. Along with the electromagnetic radiation of celestial bodies, an almost isotropic radio emission with a "thermal" spectrum spreading uniformly in all directions is recorded. This emission is called relic.

It can be hypothesized that the same galaxies, galactic clusters, and other cosmic objects have a different impact on the structuring of thinking of different people. Quite probably, the influence of objects of the observed Universe on thinking is extremely small. Nevertheless, the neurophysiological features of the human brain in combination with nonlocal quantum, as well as electrodynamics effects, can make this influence perceptible.

Theoretical analysis makes it possible to reproduce, with great certainty, the scenario of the evolution of the Universe, starting with the size of the Planck length. At the very initial moment of the origin of the Universe, all particles were collected almost at one point. Despite the subsequent expansion of the Universe, this moment was memorized in a certain way and now manifests itself, e.g., by the quantum effect of nonlocality of particle's behavior.

At the birth of the Universe, hot plasma appeared, containing particles and antiparticles. In the process of expansion and cooling of the plasma, matter was annihilated; nuclear reactions were carried out with the release of energy, photons, protons, neutrons, electrons, as well as other particles and fields. Nuclear reactions determined the composition of matter, including electrons that merged nuclei to form atoms. Hydrogen, helium, and deuterium appeared accordingly. Under the action of gravity, the created mass of particles was accompanied by the formation of inhomogeneities.

Still, there are no answers to many questions: the nature of weak interactions of particles and fields, the characteristics of initial perturbations of particles and fields, the expansion rate of the Universe, the density of cosmic matter, the nature of dark matter, etc. Perhaps the research of such phenomena will accelerate if the cognitive processes themselves, the models of human consciousness, and thinking are included in the subject of their study.

When examining the relic Universe, one deals with very high energies of particles and specific fields that have not yet been observed on the Earth. Such information cannot be obtained directly from an Earth-based experiment. At the origin of the Universe, an excess of baryons arose, which led to the absence of symmetry in the Universe: there is much more matter in it than antimatter. As a result, a hypothesis of negative pressure appeared to explain the fact of the expansion of the Universe. In turn, this hypothesis can be associated with the expansion of consciousness, which provides the emergence of insights and explains the phenomenon of generating brilliant ideas "from nowhere." In this context, an association arises that improves understanding of such a mysterious thought phenomenon as a human's ability to make correct and simultaneously unreasonable decisions.

The source of information about the Universe is the relic radiation and its fluctuations. The fluctuations are very small, and their experimental study on the Earth is impossible. They carry information about the Universe of the era that preceded the formation of galaxies. It can be used for judging the events that took place near the Planck time of the origin of the Universe (10^{-43} s). Similarly, the source of human thoughts can be recombinations of elements of the subatomic level, which are interconnected with the elements of the Universe.

The answers to questions regarding the origin and evolution of the Universe, the formation and destruction of galaxies, the ratio of the observed and dark matter, etc., depend on the chosen theory and research methods. Investigations are now primarily physical and mathematical in nature. The research base is the experimental search for rare processes, the search for new particles, the study of hypothetical particles in underground and deep-sea experiments, and the detection of gravitational waves. So far, the only way of experimental research is to obtain information from the cosmos, from LHC, as well as to conduct computational experiments. Telescopes, accelerators, supercomputers, AI, and big data analytics are now being used for this.

This cosmomicrophysical approach is accompanied by natural contradictions and limitations. The effect of nonlocality contradicts special relativity. The Earth-based particle accelerators have insufficient energy: as noted above, the deficit is 10^{12} orders of magnitude. At the same time, Earth-based accelerators provide a higher intensity of particle fluxes than cosmic sources. The optical signals coming from the cosmos are subject to very large distortions in the atmosphere and optical elements of telescopes, etc.

Perhaps, for answering the current questions in such conditions, the cosmomicrophysical research tools should be diversified and made even more interdisciplinary. An integrating role can be played, e.g., by research in the area of cognitive sciences, ideas about the nature of free will, thinking, emotions, and the transcendental states of consciousness.

References

1. Gigerenzer, G.: Gut Feelings. The Intelligence of the Unconscious. Viking, London (2007)
2. Brooks, M.: The Quantum Astrologer's Handbook. SCRIBE, Melbourne, London, UK (2017)
3. Sakharov, A.D.: Kosmomikrofizika—mezhdistsiplinarnaya problema (Cosmomicrophysics is an interdisciplinary science). Bulletin of the USSR Academy of Sciences, 4 (In Russian, Moscow, 1989)
4. Visser, M.: Sakharov's Induced Gravity: a Modern Perspective. Mod. Phys. Lett. A **17**, 977–991 (2002). https://doi.org/10.1142/S0217732302006886
5. Sazhina, O.S., et al.: Optical Analysis of a CMB Cosmic String Candidate. Advance Access Publication, MNRAS **485**, 1876–1885 (2019). https://doi.org/10.1093/mnras/stz527

Chapter 7
Nonlocal Brainstorming

In practice, the cognitive semantics of AI are especially evident in the case of network collective decision-making when all group members discussing some problem are geographically distributed. Electronic brainstorming (EB) is an example of such processes; see [1]. Indirect contact and the impossibility of taking into account the thoughts of all participants significantly complicate the achievement of mutual understanding or agreement. A common requirement is to cover the discourse of the discussed problem as holistically as possible [2]. The more difficult the problem is, the deeper details will be needed to investigate.

Numerous studies have revealed the restrictions of such distributed decision-making processes. EB can be divergent and convergent [3]. The former aims to generate new ideas, as many as possible; the latter ensures the processes of reaching consensus on the solution of problems. Due to the network character of communication, the participants of EB have restricted capabilities to understand each other. In both cases, it is useful to adopt special AI technologies that will reduce or eliminate these restrictions completely. Quite obviously, these technologies have to take into account the semantics of AI models that cannot be represented in a direct logical way; see Chap. 4.

As was shown in [4], EB has not yet become a widely used idea generation technology. According to the theoretical arguments and empirical evidence of this paper, EB is not as effective as verbal face-to-face brainstorming for group well-being and member support.

Some groups using EB generate more unique ideas than groups using nominal (verbal, face-to-face) group brainstorming, whereas others do not [5]. There are two factors through which group size may affect the efficiency of brainstorming: synergy and social loafing. Time effects were discovered as follows: nominal brainstorming groups generate more ideas than the EB ones in the first time period and fewer ideas in the last time period; the synergy from the ideas is only important when groups brainstorm for a longer time; the duration of EB should be at least 30 min.

© The Author(s), under exclusive license to Springer Nature Singapore Pte Ltd. 2021   31
A. Raikov, *Cognitive Semantics of Artificial Intelligence: A New Perspective*,
SpringerBriefs in Computational Intelligence,
https://doi.org/10.1007/978-981-33-6750-0_7

In the paper [6], the proposition that EB has weak superiority over nominal brainstorming was confirmed. The process gain vs. process loss advantages of EB may not be significant enough for enabling EB groups to outperform nominal brainstorming groups. An experiment designed to compare four brainstorming technologies (nominal, EB-anonymous, EB-non-anonymous, and verbal) showed that the groups using nominal brainstorming significantly outperformed the ones using the other three brainstorming approaches. The process losses inherent to EB impair its efficiency.

The openness of participants and their thorough attention to ideas suggested by peers are important for their creativity; see [7]. According to regression analyses, the participants of an EB process characterized by high openness were more creative, but only when they demonstrated more attention to peers' ideas; EB can be useful for enhancing the creativity of some participants.

In the paper [8], the relationship between the number of generated ideas and the time taken while groups are performing a single EB task was studied empirically. According to the results presented therein, group participation increases from the beginning, then decreases; the number of ideas generated decreases over time, but a few peaks may appear. In the paper [9], an EB system with gamification elements was described by conducting several 30-min experimental EB sessions. As was shown therein, a gamification element is the most efficient catalyst for improving the quality of ideas in terms of fluency, flexibility, and originality. In addition, no significant difference was found between the efficiency of cooperative and noncooperative (competitive) gamification.

Research literature indicates the strengths of EB over face-to-face work; for details, see [10]. The special EB application prototype helps participants by structuring the generative group process to sprint through a creative process, from problem statement to solution. As was noted in the paper [11], EB is a most intensively studied topic in the fields of information systems and computer-mediated communication. The authors attracted attention to the presentation of participants and their content, which includes spatial partitioning and marking for identifiability or anonymity, as well as some related social, affective, and cognitive aspects of the user interface.

Therefore, the main way to improve collective search methods for solutions when implementing targeted communications, brainstorming, or attempts to achieve insights (the "Eureka" effect [12]), is experimental-heuristic. The group insight can appear as the result of a process that evolves something like this:

- a long search for a good idea;
- little progress in the long search of an adequate cue;
- a precipitating event that gives a chance to get an idea;
- a cognitive snap, the time instant when an idea comes suddenly;
- generation of all details of the decision.

For example, during his famous round-the-world voyage on the HMS Beagle, Darwin was writing thoughts about evolution. But the idea was not coming, and he couldn't realize the key reasons for evolutionary processes. One day, reading Malthus's essay about the exponential expansion of the human population, he achieved an insight. He understood that organisms pass on their traits to offspring for surviving. Darwin quickly resolved a puzzle that he couldn't make for a long time. He didn't recognize the genius of his discovery immediately at once. It was a cognitive snap. Then the transformation of the awareness period came. Everybody can find many similar examples in the literature devoted to breakthrough thinking. Discoveries require a while to grow their potential.

Different methods of information structuring were suggested for accelerating collective breakthrough thinking and making the results more convergent. In the papers [13, 14], it was suggested to use inverse problem-solving methods on topological spaces and cognitive modeling, respectively. For ensuring the convergence of collective thinking processes to the goals, category theory was employed, and the concept of a "convergent monad" was introduced in [15].

Most of the modern research works proceed from indirect statements about the cognitive or unformalized semantics. The individual or collective participation of the subject makes these semantics an obligatory component of decision-support systems. A decision-making process is dead without a human, and participants need smart targeted tips from an AI system when making their decisions. The tips have to be represented in a logical and reasoned way. In these circumstances, the following supposition seems quite natural: the holistic discourse of problem-solving with human participation can be embraced within indirect approaches and, consequently, cognitive semantics, which can possibly be raised onto relativistic height or plunge to atomic depth; see Chaps. 6, 13, and 21.

The problem statement can be seen as establishing a cognitive bridge between the solid logical pillar (on the one side) and an unformalized phenomenon (on the other). This bridge cannot be built from logical structures only and requires indirect forms of descriptions. Apparently, the description in the form of knowledge base, ontologies, and even deep neural networks, cannot be attributed to such (indirect) forms.

The attempts to formalize psychological concepts such as the unconscious, archetypes, emotions, etc., always arouse the wary perception of experts. The transcendental aspect of consciousness, most likely, goes into even deeper spheres falling out of the modern scientific paradigm, perhaps, except for the philosophical and psychological ones. In this context, recall the transcendental apperception of Kant, the phenomenology of Husserl, the cognitive spaces of Lewin, or the neurolinguistic programming of Erickson.

The indirect forms representing the phenomenological processes of the transcendental world of consciousness and thoughts (see above) can be realized using cosmological approaches, quantum fields, electromagnetic waves, etc. Scientific methods do not deny the possibility of (and even the need for) creative deviations from traditional scientific postulates, not to mention the impossibility of generating something new without emotions and experiences. It expands the spectrum of scientific laws and regularities under study. Semiotics and hermeneutics are inconceivable without such deviations. Positive examples of such activities include Heyting's mathematical intuitionism and the fascination of DNA researchers with the romantic poetry of Keats. Recall the speech of Einstein when he received the Nobel Prize in Physics

(1921), in which he noted that there is no logical path to the physics laws; only intuition and resting on sympathetic understanding of experience can reach them.

Quantum systems are immersing in an infinite-dimensional Hilbert space; see Chap. 10. This space is constructed using observable parameters as a kind of AI model's product. When the study of a quantum system begins, these parameters are not associated with the operators of this space. The connection between the parameters and the operators is found later, as soon as space is constructed. Further, the process moves towards operating with formal parameters. Such a formalized operation seems somewhat impressive from the scientific viewpoint and can be useful for deterministic systems, but it is of little assistance for immersing in a hidden phenomenon, e.g., the one associated with the collective unconscious.

For this immersion, the concept of locality in a quantum context can be adopted. The locality reduces to the fact that at separate points of space, some fields should be commutative whereas others anticommutative. If the elementary part of a field means some thought of a group member, then the operations over the thoughts in the space can be used for representing the creative processes of the group. Meanwhile, such representations can create significant obstacles to the modeling of meditative phenomena since it establishes a priori restrictions in the form of the axiomatics of Hilbert spaces for phenomena that cannot be represented in a formalized way.

A reference, or even starting, point in the study of the meditative level of consciousness and its emanations could be the fact that consciousness is affected not only by energy and, possibly, electromagnetic, audio, and thermal fields of biological elements but also by the fields of the surrounding space, both nearby and at considerable distances. One cannot reject, e.g., the possibility of nonlocal quantum–mechanical (Chap. 13) and wave (Chap. 12) external physical influences on the meditative processes.

Researches into the interaction of electromagnetic fields with the cells of living organisms are of interest. Over the years, an enormous number of investigations have been carried out on the biological effects arising in the medium of low-temperature gas-discharge plasma. Nonequilibrium plasma has the ability to influence biological processes through electron energy, which is much higher than that of ions and neutral particles formed in the gas phase. Electrons interacting with a gas initiate the processes of dissociation, excitation, and ionization. However, the gas-discharge plasma itself is in contact only with the surface of a biological object; at the same time, incoherent plasma radiation can penetrate into a biologically active object. The effect of the low-temperature gas-discharge plasma on bactericidal and cytotoxic processes was studied, e.g., in [16].

Note that when conducting such studies, the interaction of electromagnetic fields with an organism, which itself may have the properties of low-temperature plasma, may be neglected. As is well known, the impact of low-temperature gas-discharge plasma radiation on the suspension of animal bone marrow cells causes a statistically significant change in the concentration of ions in the extracellular environment. Focusing on the manifestation of meditative effects on mental processes can guide these studies in the appropriate direction. Meditation is characterized by the effect of explosive deep energy activity, something like latent emotional arousal, which occurs

Fig. 7.1 Negative resistance of gas discharge gap

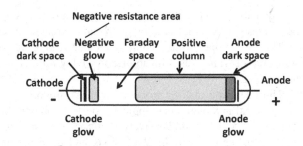

when entering the state of deep meditation. After leaving the state of meditation, this activity may sharply improve the mood, increase blood pressure, etc.

From the viewpoint of modeling, this possibly means the effect of generating radiation and electromagnetic waves, which arises due to a negative resistance during the interaction of an electromagnetic field with the cells of the body, provided that they have the properties of the low-temperature plasma. A possible appearance of similar negative resistance in a low-temperature plasma was demonstrated by the author of this book long ago when studying the nature of a gas-discharge low-temperature (plasma) analog of a microwave avalanche-transit diode (up to 26 GHz). In the dark near-cathode region of the glow discharge, an avalanche multiplication of electrons occurs; in the negative cathode glow, they have drift [17]. As was theoretically shown and experimentally confirmed in that paper, when the signal frequency decreases below a certain threshold, a negative resistance of the gas-discharge gap, significant by absolute value, can be expected; see Fig. 8.1.

Most likely, the results obtained cannot be taken for granted: they are applicable to simulate nonlocally the deep processes running in the minds of EB participants. The cognitive semantics of AI models should perhaps consider the effect of radiation generation by biological cells interacting with an electromagnetic field and low-temperature plasma Fig. 7.1.

Such interactions can play a significant role in the formation of a meditative effect in the consciousness of a participant. But only an experienced human, mastering the techniques of deep immersion in meditation and slowly going through the multi-layered path of immersion—from the consistent awareness of the external environment, body, thoughts, and feelings—can achieve the real meditative state. This pathway can be influenced by all types of biological activity, including those based on the interaction of low-temperature gas-discharge plasma and electromagnetic field with the human body.

Within the quantum theory, researchers quantize electromagnetic field; see Chap. 21. At the same time, the quantization-based formalisms of this field are a restriction on the way towards creating AI tools to model the processes of thinking and consciousness.

The main types of biological activity of living organisms with thermal effects occurring within a region of the electromagnetic field spectrum where the energy of a quantum exceeds the kinetic energy of molecules at the temperature of living organisms (from the infrared range to gamma radiation) have been well studied.

This gives hope for incorporating the obtained results into the nonlocal cognitive semantics of AI models, with application to high-quality decision-making processes, including EB.

Thus, decision-making processes with AI support have to take into consideration the nonlocal and phenomenological aspects of human thinking and consciousness. They cannot be comprehensively represented in any logical form or even using any neural networks. Indirect methods and deep biological, quantum, and other approaches are required to describe the cognitive features of decision-making with AI support. This is especially important during collective goal-setting due to the ill-defined character of the process.

References

1. Gubanov, D., Korgin, N., Novikov, D., Raikov, A.: E-Expertise: Modern Collective Intelligence. Springer, Series: Studies in Computational Intelligence, vol. 558, XVIII (2014). https://doi.org/10.1093/mnras/stz527 https://doi.org/10.1007/978-3-319-06770-4
2. Raikov, A.N.: Holistic Discourse in the Network Cognitive Modeling. J. Math. Fundam. Sci. **3**, 519–530 (2013)
3. Klimenko, S., Raikov, A.: Virtual Brainstorming. In: Proceedings of the International Scientific-Practical Conference on Expert Community Organization in the Field of Education, Science and Technologies, Triest, Italy, pp. 181–185 (2013)
4. Dennis, A., Reinicke, B.: Beta Versus VHS and the Acceptance of Electronic Brainstorming Technology. MIS Quarterly **28**(1), 1–20 (2004). https://doi.org/10.2307/25148622
5. Dennis, A.R., et al.: Patterns in Electronic Brainstorming. Int. J. e-Collaboration (IJeC) 1(4), 20 (2005). https://doi.org/10.4018/jec.2005100103
6. Pinsonneault, A., et al.: Electronic Brainstorming: the Illusion of Productivity. Inform. Syst. Res. **10**(2), 110–133 (1999). https://doi.org/10.1287/isre.10.2.110
7. Pi, Z., et al.: The Relation Between Openness and Creativity is Moderated by Attention to Peers' Ideas in Electronic Brainstorming. Interact. Learn. Environ. (2019). https://doi.org/10.1080/10494820.2019.1655458
8. Deng, H., Zhang P., Sun, J.: An Empirical Study on Dynamic Process of Electronic Brainstorming. In: Proceeding of 4th International Conference on Wireless Communications, Networking and Mobile Computing, Dalian, pp. 1–4, IEEE (2008). https://doi.org/10.1109/WiCom.2008.2753
9. Yuizono, T., Xing, Q., Furukawa H.: Effects of Gamification on Electronic Brainstorming Systems. In: Yuizono, T., Zurita, G., Baloian, N., Inoue, T., Ogata, H. (Eds.), Collaboration Technologies and Social Computing. Communications in Computer and Information Science, vol. 460. Springer, Berlin, Heidelberg. (2014). https://doi.org/https://doi.org/10.1007/978-3-662-44651-5_5
10. Liikkanen, L.A., et al.: Next Step in Electronic Brainstorming: Collaborative Creativity with the Web. In: Proceedings of the International Conference on Human Factors in Computing Systems (CHI 2011), Vancouver, Canada, pp. 2029–2034 (2011). https://doi.org/10.1145/1979742.1979929
11. Zelchenko, P., Ivanov, A., Mileva E.: Reviewing the Interaction Aspects of a Line of Electronic Brainstorming Social Interfaces. In: Luo, Y. (Ed.) Cooperative Design, Visualization, and Engineering (CDVE 2018). Lecture Notes in Computer Science, vol. 11151. Springer, Cham (2018). https://doi.org/10.1007/978-3-030-00560-3_45

12. Perkins, D.: The Eureka Effect. The Art and Logic of Breakthrough Thinking. Norton, New York, London (2001)
13. Raikov, A.N.: Convergent Cognitype for Speeding-up the Strategic Conversation. In: IFAC Proceedings Volumes, vol. 41(2), pp. 8103–8108. Seoul, South Korea (2008). https://doi.org/10.3182/20080706-5-KR-1001.01368
14. Raikov, A.N., Panfilov, S.A.: Convergent Decision Support System with Genetic Algorithms and Cognitive Simulation. In: Proceedings of the IFAC Conference on Manufacturing Modelling, Management, and Control, Saint Petersburg, Russia, pp. 1142–1147 (2013). https://doi.org/10.3182/20130619-3-RU-3018.00404
15. Raikov, A.N., Ermakov, A.N., Merkulov, A.A.: Assessments of the Economic Sectors Needs in Digital Technologies. Lobachevskii J. Math. **40**(11), 1837–1847. Pleiades Publishing, Ltd (2019). https://doi.org/10.1134/S1995080219110246
16. Kieft, E., Darios, D., Roks, A.J.M., Stoffels, E.: Plasma Treatment of Mammalian Vascular Cells: A Quantitative Description. IEEE Trans. Plasma Sci. **33**(2), 771–775 (2005)
17. Lobov, G.D., Raikov, A.N.: Possibilities of Creating a Gas-Discharge Analogue of a Superhigh-Frequency Avalanche-Transit Diode. In: Proceedings of the Moscow Power Engineering Institute. Zhukov, V.P. (Ed.), Moscow, Issue 231, pp. 71–74 (in Russian) (1975)

Chapter 8
Purposeful Creativity

Goal-setting cannot be completely described in a formalized way. A goal is the anticipation of a process or its result in the human mind. First of all, the complexity of the goal-setting process is due to the influence of a subjective factor on it. This complexity can be qualitatively characterized by considering emotions, the individual and collective unconscious, or using mathematics and physics. For example, a very scientific explication of the concept of complexity appears in attempts to elaborate a Grand Unified Theory (GUT). When defining complexity, one should take into account history, ethics, and culture.

The crisis of economics, the subtlety of diplomacy, and the confusion of politics form the external context of goal-setting. There may exist no causal chain of implications in reasoning: people often make decisions that are completely correct but have no explanation. In this case, "correct scientific goal-setting" becomes a task completely inaccessible for a human.

To reduce the complexity of goal-setting, modeling is often used: an immense situation is replaced by the logic of its model. However, goal-setting requires the direct involvement of a human in the process. Then only some part of the effort is allocated to the actual logic of goal-setting, and thoughts and emotions will replace most of the process.

In many cases, goal-setting cannot be implemented by extrapolating the previous dynamics of events, e.g., using statistical analysis. The goals are often strategic, i.e., ambitious: they do not lie on the inertial curve. Goals are needed; otherwise, the development of the situation will put at high risk the existence of a human organization, lead to a crisis, bankruptcy, etc. The choice of an appropriate approach to goal-setting predetermines the methods of modeling and solving problems; they can be direct and inverse, considered in metric or nonmetric spaces. The situation of goal-setting has been illustrated above; see Fig. 3.1.

Imagine that on a sunny day, you are standing by the lake and admiring nature. A beautiful scene, transparent air, a smell of soil and grass, and a smooth surface of warm water—everything makes the moment unforgettable. You are engrossed

© The Author(s), under exclusive license to Springer Nature Singapore Pte Ltd. 2021
A. Raikov, *Cognitive Semantics of Artificial Intelligence: A New Perspective*,
SpringerBriefs in Computational Intelligence,
https://doi.org/10.1007/978-981-33-6750-0_8

by emotions. You can take a photograph of the scene, and in the evening with your family, you can view these unforgettable moments on the computer. Beauty will fade, the smell will not be the same, and the Sun will not warm. Emotions will weaken.

However, family members know and feel the richness of the term "water." Water accompanies everyone from birth. Studies have filled this word with both magical content and the laconic formula H_2O. Summer is associated with the river and the sea. We are made of water, and there are ways to purify it. Many riddles of water have not yet been solved. Water is rain and downpour, thunder and lightning, and hazard and silence. Water is life felt with the entire soul. Water is a primary criterion for life on another planet. Many of these shades of water are instantly perceived by a human. This feeling enhances the experience of viewing photos. However, a computer would hardly use this knowledge and feelings for the logical derivation of decisions in the course of goal-setting.

A computer sees the picture, an outer wrapper of the infinite in a digital representation. When processing soulless data by a computer, one tries to fill the deficit of human feeling using logical ontologies, pictures, sounds, and descriptions. As has been noted above, this is the so-called denotative semantics; see Chap. 4. It has a nature of projection: some signs are mapped into other sign structures. However, as a result of this procedure, a computer does not acquire any feelings. In other words, much is lost when processing data using denotative semantics.

A human is able to achieve insight; see Chap. 7. On the one hand, insight is treated as a characteristic of human thinking, when a decision is reached by comprehending the whole, and not only as a result of the analysis. On the other hand, insight is used to describe a phenomenon when a human experiences sudden enlightenment or unreasonable clarification after a series of fruitless attempts to find a solution [1].

A proven scientific way to accelerate problem-solving in any complex situation, including creative ones, is to project an image of the situation into visual, algebraic, thermodynamic, logical, quantum–mechanical, and other structures. In this context, goal-setting can be represented as a process of forming a group and reaching a collective agreement on goals and ways of their achievement among group members. There may exist several goals. The choice of an appropriate way to achieve goals includes the selection and assessment of resources (material, intellectual, financial, etc.), the development of an action plan, and the ordering and optimization of actions execution. With such an approach, the goal-setting problem can be formulated in mathematical terms as follows; also, see Chap. 18.

Problem. Let $y_0 \in Y$ be an exact goal (its name and value, or indicator). Assume that under a mapping $A{:}X \to Y$, it has a unique preimage x_0 in X (an exact set of resources and actions to achieve this goal, a plan). The problem is to associate with each element V_δ (an approximate value of the goal) in a filter (a collection of sets with a non-empty intersection) of all neighborhoods of the point y_0 (the set of inexact values of the goal) a point $x_\delta \in X$ (an inexact plan, action, or resource) such that $x_\delta \to x_0$ (the plan is refined in a convergent way, representing a more and more accurate sequence of actions). Each x_δ (inexact plan, action, or resource) with this property will be called an approximate solution of the equation $Ax = y$ corresponding to the neighborhood V_δ of the point y_0.

This problem can be solved on metric spaces (when a metric is used to estimate the distance between points) and on topological spaces (when the distance between points is determined through the intersection of their neighborhoods).

When setting strategic goals, the problem can become inverse, without any representation in metric spaces. The solution of such a problem can be represented in a topological space and characterized by ill-posedness, no longer matching the definition below; also, see [2].

Definition 1 Let $\{V_\delta\}$ be a filter of all neighborhoods of a point y_0. The equation $Ax = y$ for $y = y_0$ is a well-posed problem if:

(1) The intersection $\cap A^{-1}V_\delta$ of all complete preimages contains only one point x_0 (existence and uniqueness);
(2) The filter in X generated by the set of complete preimages $A^{-1}V_\delta$ converges to x_0 (the continuous dependence of the solution on the initial conditions).

Condition (1) ensures that (a) the problem has a solution of the form $x_0 = \cap A^{-1}V_\delta$; and (b) this solution is the only correct one (adequate to the real situation and suits all participants). According to condition (2), the resources.

$(A^{-1}V_\delta)$ suffice for achieving the goals (y_0), and there are different alternatives to use the resources (a filter in X); it remains to find an appropriate sequence of their use. The solution can be found by enumerating all alternatives or ranking them by multiple criteria; linear or nonlinear programming, genetic algorithms, or some other methods can be employed as well.

In this case, the point y_0 (the exact goal, its name and value, or indicator) is not necessarily an inner point of the codomain of the operator A. Therefore, arbitrarily close to the exact value of the goal, there may exist points y without preimages in X; as a result, the problem may have no solution. This remark is very important: the goal setter should be disillusioned that small changes in the values of goals will correspond to small changes in action plans. In incorrect (unstable) problems, a slight change in the initial conditions can lead to a fundamental change in the way of achieving the goal. In other words, the set of preimages $A^{-1}V_\delta$ of all neighborhoods of the point y_0 has a single intersection point x_0, thereby being a basis of the filter $\{A^{-1}V_\delta\}$; but the convergence of this filter to the point x_0 is not required. A problem with these properties is said to be unstable.

A way to eliminate this obstacle, as well as to implement a purposeful and convergent search for a solution [2], is to contract the mapping A to a compact set M of the space X: provided that the *graph* of the mapping A is *closed* (see Definition 2 below), this will guarantee the convergence of the solution. A topological space is compact if and only if a finite subcover can be selected from each of its open covers [3]; also, see Chap. 18 and Fig. 18.1.

The closed sets $A^{-1}V_\delta$ are needed. They fully (holistically) interpret the available resources for achieving the goal y_0. For this purpose, the concept of a family of centered sets is also required. A family of sets is such if and only if the intersection of any finite set of elements from this family is non-empty. A relationship can be established between the concepts of a centered system and compactness as follows.

Theorem 1 *A topological space is compact if and only if each centered system of closed sets in it has a non-empty intersection.*

A proof using de Morgan's formulas was presented in the book [3]. Also, it can be demonstrated that each compact subset of the Hausdorff space is closed. A mapping with a closed graph is defined as follows.

Definition 2 A mapping $A: X \to Y$ is called a mapping with a closed graph if from the conditions that (a) the filter $\{E_\alpha\}$ on the domain $D(A)$ converges in X to $\bar{x} \in X$ and (b) the filter induced by the sets $AE_\alpha \subset Y$ converges to $\bar{y} \in Y$, it follows that $\bar{x} \in D(A)$ and $y = A\bar{x}$.

Under a mapping with a closed graph $A: X \to Y$, the image of a compact set is closed. This fact follows from a theorem proved in [2]. The compactness of the set implies the need to structure all means of achieving the goals into a finite and observable number of components $A^{-1}V_\delta$, each of which corresponds to its own subgoal or inexact goal V_δ. Only the presence of such a morphism and the convergence of the corresponding filters on the spaces X and Y guarantee that the goal y_0 will be achieved. Thus, the closedness of the graph of a mapping requires establishing a correspondence between the goals and means of achieving them.

The creative search for solutions and ideas (e.g., in the course of brainstorming, etc.) may include the following aspects [1]:

- The search is characterized by many directions with a small number of solutions (immenseness).
- The search runs in space without any clues or hints regarding a correct course of action or decision.
- The search for a solution closes within some part of the problem, in which (as it turns out later) there is actually no solution.
- Deceptive promises to get the right solution are encountered along the way of problem-solving.
- When studying the issue of reaching a group agreement in goal-setting, the following elements of space should be emphasized:

  - the space Y of goals, which contains the exact goal y_0 and the set of inexact goals indexed by δ;
  - the space X of resources and means of achieving goals, which contains the exact solution x_0;
  - the mapping operator $A: X \to Y$.

In this case, the domain of the operator A can be reduced to a subset F of the set X, and the range of the operator A to some subset of the set Y. Applied research shows that when solving ill-posed problems, it is fruitful to extend the well-known regularization method to derive necessary conditions of a corresponding homeomorphism (a continuous bijection); also, see Chap. 18.

When modeling a problem using AI tools and methods for solving ill-posed problems in topological spaces, the elements of the model have to match the cognitive semantics of AI, actually carried by the participants in the problem-solving process. They bring qualitative (non-formalizable) information into the problem-solving process, which requires a kind of cognitive architecture that establishes a bridge between the formalized AI model and informal human thinking.

References

1. Perkins, D.: The Eureka Effect. The Art and Logic of Breakthrough Thinking. Norton, New York, London (2001)
2. Ivanov, V.K.: Incorrect Problems in Topological Spaces. Siberian Math. J. **10**, 785–791, (Novosibirsk, Russia, 1969)
3. Kelley, J.L.: General Topology, Springer, New York, XIV, 298 p. (1975)

Chapter 9
Cognitive Structures

The ultimate goal of research into cognitive architectures is to model the human mind, eventually enabling to build human-level AI. Some criteria of cognitive architectures include understanding, free will, flexible behavior, real-time operation, rationality, large knowledge base, learning, development, linguistic abilities, self-awareness, and brain realization. Investigations in this area of science cover ecological and evolutionary realism, adaptation, modularity, routineness, and synergistic interaction [1]. Also, an uncertainty regarding dichotomies (implicit/explicit, procedural/declarative, etc.), the modularity of cognition, and the structure of memory are considered.

The core cognitive abilities, such as perception, attention, and action choice, as well as memory, learning, reasoning, and meta-reasoning, were discussed in [2]. About three hundred cognitive architectures describe these abilities. More than 2500 relevant publications were analyzed, and 900 projects were implemented in practice and reviewed.

In the paper [3], some issues concerning the broad areas of cognitive architectures, such as perception, memory, attention, actuation, social interaction, planning, motivation, emotion, etc., were discussed.

ACT-R, Soar, and Sigma are well-known systems with a different structural organization and different approaches to the modeling of human cognitive abilities; see [4]. A new standard model of the mind was proposed therein as a consensus among different architectures. ACT-R is a hybrid cognitive architecture. It consists of some logical mechanisms that can be used to predict and explain human behavior, including cognition and interaction with the environment. A theory of cognition is realized as a computer program. The tools for working with ACT-R take into account emotions and physiology, increasing usability [5]. It can be described as a way of specifying how the brain is organized for enabling individual processing modules to produce cognition.

Soar is another system providing the interaction between procedural memory (knowledge about ways of doing things) and working memory (knowledge about the current situation) to support the selection and application of operators. Working

© The Author(s), under exclusive license to Springer Nature Singapore Pte Ltd. 2021   45
A. Raikov, *Cognitive Semantics of Artificial Intelligence: A New Perspective*,
SpringerBriefs in Computational Intelligence,
https://doi.org/10.1007/978-981-33-6750-0_9

memory consists of symbolic graph structures. The procedural memory embraces a set of "if–then" rules that are matched against the contents of working memory. Soar selects applications of operators, which are generated, evaluated, and applied by rules.

Achieving Artificial General Intelligence (AGI) is the goal of a few architectures: in addition to Soar and ACT-R mentioned above, these are NARS [6], LIDA [7], SiMA [8], Sigma [9], and CogPrime [10]. There are many research works devoted to particular aspects of cognition. For example, ARCADIA [11] and STAR [12] take into account human attention; CELTS [13], emotions; Cognitive Symmetry Engine [14], the perception of symmetry.

NARS is the non-axiomatic reasoning system that processes natural language by reasoning and learning. It is a formal model of an AGI system, attempting to provide a unified theory, and an AI system as well. Limited logical resources are a part of the NARS definition. There exist many problems that cannot be solved by logic only. As an example, consider the problem arising in space exploration with a radio telescope: there is no absolutely accurate method to recover the original analogue information about space object, which was transformed by receiving detector in a digital form. Also, note many other phenomena, such as the decision-making problem, where an axiomatic logic cannot support problem-solving, or the implication paradox, where irrelevant data and uncaused insight can yield correct results.

LIDA is a learning intelligent distribution agent of cognitive architecture. This artificial cognitive system simulates a broad spectrum of cognition, from low-level perception to high-level reasoning. The LIDA architecture proceeds from the idea that human cognitive functions are realized via frequently repeated (~10 Hz) interactions, called cognitive cycles, between conscious contents, various memory systems, and action selection. These cognitive cycles are composed of higher-level cognitive processes. The LIDA architecture employs several modules that are designed using computational mechanisms. The LIDA cognitive cycle is divided into three phases: the understanding phase, the attention (consciousness) phase, and the action selection and learning phase.

The understanding phase is activated by feature detectors in sensory memory. The output is obtained using perceptual associative memory and consists of more abstract entities, such as objects, categories, actions, events, etc. The local associations are combined with the perception of the environment to generate a current situational model of what is going on around. The attention phase forms coalitions of the current situational model, which subsequently compete with one another for attention (i.e., for getting a place in the conscious contents). These contents then broadcast, initiating the action selection phase. New associations and the reinforcement of old ones occur as the conscious broadcast reaches various forms of memory, perceptual, episodic, and procedural. The selected behavior triggers memory to produce a suitable algorithm for its execution, which completes the cognitive cycle.

In the review [2] mentioned above, cognitive architectures were classified. According to this classification, there exist three main classes of methods: emergent, hybrid, and symbolic. All these methods rest on logic. The symbols (labels, strings of characters, frames), production rules, and non-probabilistic logical inference are symbolic, but neural networks are sub-symbolic. Hybrid representation may combine elements from both representations, symbolic and sub-symbolic.

For a long time, the vision was viewed as the dominating human sensory modality. During recent decades, the vision in human sensory experience has been considered in a strong connection with cognitive understanding. Visual processing may include the stage of object recognition; in this case, some meaning is assigned to objects using available knowledge. Attention mechanisms and emotions also affect all stages of visual processing; see [15].

Internal cognitive factors do not determine the behavior directly, rather biasing the selection. They can be considered short-term, long-term, and life-long factors loosely corresponding to emotions, drives, and personality traits in humans. Emotions in cognitive architectures are typically modeled by transient states (anger, fear, joy, etc.) that influence cognitive abilities. They can modify the global plan and final selection. Plans that confront a threat have higher utility when morale is high but lower utility when fear is high. Stress affects decision-making, and emotions affect blackjack strategy; reasoning depends on the state of anxiety; and so on, see [16].

As can be observed, the dominant approach in the design of cognitive architectures involves symbolic and logical tools. At the same time, human thinking has characteristic features that can be given a very limited description in logical terms. These features are the following:

- The order of solution choice determines the way of achieving the goal (*noncommutativity*).
- The search for solutions is carried out sequentially, but operations can be combined or parallelized in an arbitrary way (*associativity*).
- The search space is infinite and, due to its size, the search space cannot be completely explored in a limited time (*infiniteness*).
- The definition and description of events depend on the observer since they are subjectively perceived (*observability*).
- The group consciousness is affected by distant events, and sometimes in a hidden way (*nonlocality*), etc.

In view of such features of thought processes, one should employ not only logical, psychological, and heuristic approaches but also physical and mathematical methods to support decisions. More attention should be paid to the construction of cognitive semantics without any direct formalized representation. Indirect methods to represent cognitive semantics can be implemented, for example, using wave theory, quantum physics, and the theory of relativity. This approach is well justified by the existence of analogies between events in the cognitive and physical world.

The analogies between the physical and cognitive environments are as follows. In quantum physics, the observer's attempts to measure a phenomenon cause its collapse, e.g., the disappearance of the interference effect. So in the cognitive case, the observer's attempts to formalize cognitive semantics (thoughts, feelings, and emotions) lead to their disappearance and transformation into the denotative ones.

As a consequence, the power of the cognitive phenomenon is lost with the formal explication of the cognitive aspects of consciousness.

In a quantum system of two particles that can be at a great distance from one another, the detection of a certain state of one quantum particle instantly determines the state of the other; see Chap. 13. Note that in this case, no transfer of information occurs. There is an obvious analogy with the phenomenon of cognitive insight, when enlightenment—after a long and fruitless reflection on the issue—comes suddenly, as if under a push from the outside; see Chap. 7.

For covering these aspects, a mathematical space of adequate abstract level containing logical and phenomenological components has to be defined. This space needs the time axis for considering the dynamic characteristics of the situation. But it is not only the time that some observer of the events can get from his or her wrist-watch. The observer affects the time and events because the communication means have electrodynamic or acoustic, i.e., wave-particle, quantum, and relativistic nature. Due to it, different fields (gravitation, electromagnetic, strong, and weak ones) must be considered.

Modern computations deal with digital data, operating in a formalized space. It is neither a cognitive space nor a wave. It is just their digital representation. Binary digits and samples have limited ability to represent signals in an adequate way. The natural continuous signals or sounds have to be repeated absolutely with covering their full spectra; otherwise, the digitalized representations of the natural signals cut or even destroy the signal spectra. According to the well-known Nyquist theorem, various limiters (e.g., windows) for the digital transformation of waves and signals literally explode high frequencies; see Chap. 11. This leads to the use of the Gaussian window (the Gabor transform), which attenuates high frequencies on both sides of the window; see [17]. Different filters intended to transform a signal for facilitating calculations, e.g., the Haar wavelet, may significantly distort the signal.

Two different components of the space located on both sides of the bridge—phenomenological and logical—are the two characteristics of each event that happens in space. They look like imaginary and real numbers used for electromagnetic wave representation. The former class of numbers describes the phase of harmonic oscillations; the latter, their amplitude. This form can be tried for representing the unconscious, feelings, and emotions, which have a phenomenological character. The irrational abilities of humans look like a mathematical function with real and imaginary numbers, where the latter is designated as i (the square root of -1). In this case, each event can be mathematically represented as $a + ib$, where a and b are real numbers whereas i denotes the imaginary unit.

For these conditions of events representation, which are characterized by the uncaused, phenomenological, quantum, and relativistic nature, complex numbers, infinite manifold, signal continuity, and samples, group theory with quantum and relativistic fields may be useful.

References

1. Sun, R.: Desiderata for Cognitive Architectures. Philos. Psychol. **17**(3), 341–373 (2004)
2. Kotseruba, I., Tsotsos, J.K.: 40 Years of Cognitive Architectures: Core Cognitive Abilities and Practical Applications. Artif. Intell. Rev. **53**(1), 7–94 (2020). https://doi.org/10.1007/s10462-018-9646-y
3. Adams, S., et al.: Mapping the Landscape of Human-Level Artificial General Intelligence. AI Mag. **33**(1), 25–42 (2012)
4. Laird, J.E., Lebiere, C., Rosenbloom, P.S.: A Standard Model for the Mind: Toward a Common Computational Framework Across Artificial Intelligence, Cognitive Science, Neuroscience, and Robotics. AI Mag. **38**(4), 3–26 (2017)
5. Ritter, F.E., Tehranchi F., Oury, J.D.: ACT-R: A Cognitive Architecture for Modeling Cognition. WIREs Cognitive Sci. **10**(3) (2019). https://doi.org/10.1002/wcs.1488
6. Wang, P.: Natural Language Processing by Reasoning and Learning. In: Proceedings of the International Conference on Artificial General Intelligence, pp. 160–169 (2013)
7. Faghihi, U., Franklin, S.: The LIDA Model as a Foundational Architecture for AGI. In: Wang, P., Goertzel, B. (Eds.), Theoretical Foundations of Artificial General Intelligence. Atlantis Thinking Machines, vol. 4. Atlantis Press, Paris, pp. 103–121 (2012)
8. Schaat, S., et al.: Interdisciplinary Development and Evaluation of Cognitive Architectures Exemplified with the SiMA Approach. In: Proceedings of the EuroAsianPacific Joint Conference on Cognitive Science, CEURWS.org, pp. 515–520 (2015)
9. Pynadath, D.V., Rosenbloom, P.S., Marsella, S.C.: Reinforcement Learning for Adaptive Theory of Mind in the Sigma Cognitive Architecture. In: Goertzel, B., Orseau, L. and Snaider, J. (Eds.), Artificial General Intelligence, Lecture Notes in Computer Science, vol. 8598. Springer, Cham, pp. 143–154 (2014). https://doi.org/10.1007/978-3-319-09274-4_14
10. Goertzel, B., Yu, G.: A Cognitive API and Its Application to AGI Intelligence Assessment. In: Goertzel, B., Orseau, L., Snaider, J., (Eds.), Artificial General Intelligence, Lecture Notes in Computer Science, vol. 8598. Springer, Cham, pp. 242–245 (2014). https://doi.org/10.1007/978-3-319-09274-4_25
11. Bridewell, W., Bello, P.F.: Incremental Object Perception in an Attention-Driven Cognitive Architecture. In: Proceedings of the 37th Annual Meeting of the Cognitive Science Society, pp. 279–284 (2015)
12. Tsotsos, J.K.: Attention and Cognition: Principles to Guide Modeling. In: Zhao, Q. (ed.) Computational and Cognitive Neuroscience of Vision, pp. 277–295. Elsevier, New York (2017)
13. Faghihi, U., Poirier, P., Larue, O.: Emotional Cognitive Architectures. In: D'Mello, S., Graesser, A., Schuller, B., Martin, J.C. (Eds.) Affective Computing and Intelligent Interaction (ACII 2011). Lecture Notes in Computer Science, vol. 6974. Springer, Berlin, Heidelberg (2011). https://doi.org/10.1007/978-3-642-24600-5_52
14. Henderson, T.C., Joshi, A.: The Cognitive Symmetry Engine. Technical Report UUCS-13-004 (2013)
15. Tsotsos, J.K.: A Computational Perspective on Visual Attention. MIT Press, Cambridge, 328 p. (2011)
16. Schiller, M.R.G., Gobet, F.R.: A Comparison Between Cognitive and AI Models of Blackjack Strategy Learning. Lect. Notes Comput. Sci., pp. 143–155 (2012)
17. Zhang, D.: Fundamentals of Image Data Mining Analysis, Features, Classification and Retrieval. Federation University Australia, Churchill, Australia. 333 p. (2019). https://doi.org/10.1007/978-3-030-17989-2

Chapter 10
Quantum Semantics

The development of quantum semantics (QS) for AI models is a fascinating area of research, which intrigues by the ability to describe the behavior of objects taking into account their internal structure by literally replacing these objects, completely duplicating them, and making their *atomic twins*. The main idea of QS consists in representing the problems and processes in a more holistic way using a quantum interpretation of the unformalized and uncertain characteristics of phenomena [1, 2].

An increased interest in quantum information and calculation has emerged in recent years; for example, see [3, 4]. The model of quantum calculation is based on a closed or isolated quantum mechanical system. This is an ideal system without any perturbing interactions with the external environment, which are referred to as decoherence. But real systems cannot be closed. Interactions with the external environment lead to information decay and quantum errors.

Roughly speaking, QS maps an AI model into quantum computation, which in turn changes the state of a given quantum system. Mathematically quantum computation can be represented by a unitary operator U. The latter is a matrix representation of a linear operator if $U^{\dagger}U = I$, where $U^{\dagger} = (U^{T})^{*}$ denotes the complex conjugate transpose of a matrix U and I is an identity operator.

QS can be useful for improving the quality of decision-making and pattern recognition; see [5, 6]. For example, in the paper [7], QS was applied for predicting the long-term needs for educational services under a high uncertainty of the market. The first step of creating QS in that paper included cognitive modeling and its denotative semantic interpretation through mapping the model into the relevant subsets of Big Data. A quantum operator can be used for predictions. As some examples, note the quantum genetic algorithm (QGA), the entanglement operator (the Oracle Algorithm), and the Hadamard operator (HO). The HO creates a superposition of all randomly chosen solutions in the form of a generalized solution space. The HO provides the correlation between the desired solutions, whereas the OGA rejects the unnecessary solutions, e.g., using the Grover quantum algorithm. Thus,

A. Raikov, *Cognitive Semantics of Artificial Intelligence: A New Perspective*,
SpringerBriefs in Computational Intelligence,
https://doi.org/10.1007/978-981-33-6750-0_10

the quantum approach can be employed to forecast the future needs for educational services, as the superposition of all possible solutions.

In recent years, researchers and engineers have become increasingly interested in using AI in the area of pattern recognition. Due to its deep learning tools, AI has significant advantages over other approaches in such applications. This aspect has been intensively investigated since the emergence of AI, but without any progress in understanding how to penetrate into an object and to grasp its internal structure described by the methods of chemistry, biology, and physics during recognition procedures. X-ray, thermal imagers, THz devices, nuclear magnetic resonance therapy [8], etc., assist in such penetration. However, this can be done from a very close range only. The problem of penetrating into the deepest level of events and objects via AI tools can be described by the following requirements:

- The AI approach should consider not only the denotative semantics (logic, images, words) of AI models but also their cognitive semantics (concepts, emotions, etc.), which cannot be interpreted in logical terms only.
- When forecasting the behavior of patterns, an AI model should take into account possible (and unpredictable) changes in the situation.
- New random factors introduced by a human into an AI model should have a meaningful and purposeful character.
- A multi-level AI control system should ensure the sustainability and purposefulness of collective decision-making.
- The elements of the AI models' semantics can change their states in a quantized manner.
- The behavior of each element of the semantics of AI models depends on the state of some subject or object, whose location and state are unknown.

A direct analysis shows that these requirements cannot be implemented using classical AI methods. However, it seems that new approaches like QS would be appropriate here. Special quantum operators can help going deep inside an object, taking into account the impact of the external environment on it; see Chap. 13.

The same idea of using quantum operators can be cultivated for the robust prediction of development scenarios with a weakly certain image of future events. The very nature of QS [7, 9] allows fulfilling the requirements in the following way:

- The behavior of quantum particles is represented in an infinite-dimensional space, in contrast to the representation of objects in a finite set of characteristics within the classical paradigm. It can help to represent the cognitive semantics of AI and elements of objects at the atomic and sub-atomic levels.
- Within the classical paradigm, the mass and size of a particle remain the same when it changes location. Within the quantum paradigm, changing the location of a particle means replacing this particle with another one. This process resembles the changing of a real object's meaning in a different environment.
- In the quantum interpretation, a particle represents both a particle and a wave. This causes the interference effect when a moving particle passes through obstacles.

This looks like proving the fact that different arrangement of the real objects give different synergetic effects.

- Within the quantum paradigm, any attempt to measure the state of a particle leads to collapse, and the interference effect disappears. This looks like trying to penetrate into the inner structure of a recognized object by the destruction of this object.
- Quantum representations involve complex numbers, which have real and imaginary components. The field of complex numbers suits well to represent and calculate wave processes and resonance effects.
- The state of a quantum particle changes in a jump-like manner, and a quantum particle cannot have zero values of some combinations of its parameters (position and momentum, energy and time), unlike a classical object, which can be converted to a point and changes its behavior smoothly.
- Each classical object can be measured autonomously, which is impossible in the quantum case due to the nonlocality effect: in real nature, there is a general relation between the states of particles.
- Within the quantum paradigm, all phenomena are random: many experiments are required to test and confirm a phenomenon; in the classical case, this looks like errors occurring in measurements.
- The behavior of a quantum particle is not deterministic, i.e., it can change states due to an unknown reason (the so-called quantum fluctuations). This looks like a human who can make uncaused, albeit correct, decisions.
- Any classical simulation requires separating an object from its model, which leads to different assumptions. The quantum approach allows for a direct consideration of phenomena as information, thereby ensuring the fusion of a real object and its models. As a result, the object is turned into the real semantics of the AI model.

With the listed features of QS, there is hope that quantum analogies will bring long-term forecasting processes closer to reality and improve their reliability. However, many issues are yet to be solved.

The consideration of AI as processes on digital computers proceeds from the assumption that nature can be represented with digital models. But such a representation is not relevant, at least because human decision-making can be incorrect and non-continuous. In recent years, researchers have become increasingly interested in quantum semantic computing systems, involving non-monotonic quantum inference for achieving a better performance; see [2, 10]. This vision requires a new method of computing in order to take quantum effects into account.

In QS, the quantum complementarity principle can be applied for increasing the level of holistic representation of the phenomena. For example, the Heisenberg Uncertainty Relation may reflect the characteristics of ill-defined objects. In this case, the quantization of an object can be described by the relation $\Delta E \Delta t \geq r$, where the variables ΔE and Δt express an intuitive understanding of the object's energy and the uncertainty of its lifetime, respectively; r is a constant (the Planck constant in quantum physics). QS can represent the behavior of the atomic components (ACs) of a recognized object, in the case when these ACs or waves emitted by ACs collide

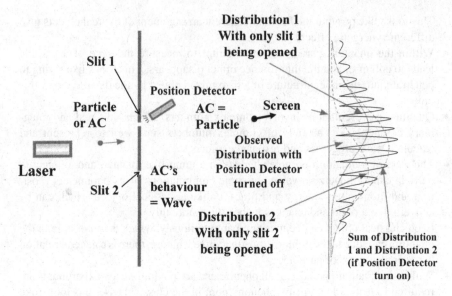

Fig. 10.1 Behavior of ACs penetrating through two slits

with obstacles. To demonstrate this, the well-known two-slit effect can be used; see Fig. 10.1. Such an analogy serves to describe the hidden characteristics of the objects.

The extending means of ACs in the information space, which is created by the interference of objects with the external environment, under certain conditions may obey the Zipf distribution law for the distribution of objects [11]. According to this empirical law, many types of data studied in different areas of knowledge can be approximated by the Zipf distribution.

The recognized object remains closed or having an opaque border until its interaction with the external environment. In this case, it is necessary to extract the hidden information from the superposition of different characteristics of ACs. In quantum formalism, the situation can be described in matrix terms.

Consider the Hadamard matrix (H) [3] and a matrix representing the state vector of the recognized object or event, including their different internal components. For example, for any object, it can be the finite number of ACs with their frequencies of occurrence f_i, where i is the rank of the ith AC. Each AC has its own state, which can be represented in the quantum denotation $|\psi\rangle = |f_i\rangle$ associated with any closed quantum system in an infinite-dimensional Hilbert function space (an inner product complex space). An inner product space is a vector space assigning a scalar value to each pair of vectors.

A Hadamard matrix is a square orthogonal matrix with entries equal to either $+1$ or -1. This matrix can be applied for creating new possible meanings of each AC by the complementary expansion, represented by f_i; see Fig. 10.2. Changing the meanings of different elements of the objects improves the understanding of

Fig. 10.2 Complementary expansion of AC

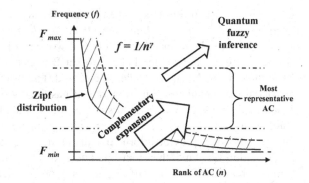

events by increasing the diversity of the AI model representation of objects under consideration in the course of their study, recognition, and processing.

Any two different rows of a Hadamard matrix are perpendicular vectors. For example, quantum fuzzy inference in Fig. 10.2 is performed by the new histogram of ACs. These ACs can have a frequency $f_i \leq F_{max}$, where i is a finite rank of an AC. It can be represented as follows: for each AC with f_i, the new meaning is derived by multiplying the matrix H on the AC state $|\psi\rangle = |f_i\rangle$:

$$H|\psi\rangle = H|f_i\rangle = 2^{1/2}\begin{bmatrix} 1 & 1 \\ 1 & -1 \end{bmatrix}\begin{pmatrix} f_i \\ 0 \end{pmatrix}. \tag{10.1}$$

This Hadamard transformation generates the states of quantum elements as the superposition of two classical states of the meanings f_i and 0. In Fig. 10.2, it can be illustrated by the expanded space of ACs and their frequencies that are below the frequency of the Zipf distribution curve. This space also contains a variety of combinations of ACs (for example, $f_1 * f_2$). It is a representation of the ACs for obtaining a higher level of holistic description during the subsequent development of cognitive semantics.

The quantum inference is fuzzy; see [12, 13]. In addition, it demonstrates well the applicability of the *entanglement* principle; also, see Chap. 13. The *entanglement* phenomenon has a connection to a fundamental property of quantum systems called non-separability (or nonlocality), which is based on the *superposition* principle and an essential characteristic of a quantum system: the behavior of its components is described by the tensor (infinite) product but not by the Cartesian (infinite) production.

Now imagine the following situation: searching for a fruitful idea (further called true), chief **C** invites two co-workers, **A** and **B**. At the beginning, both simultaneously generate three ideas; however, co-worker **A** is not interested in co-worker **B** to recognize the true idea, as they are opponents. Co-worker **A** prefers co-worker **B** to make a mistake. At the same time, co-worker **B** may be aware of it and, showing no sign of it, will try to recognize the true idea known to co-worker **A**.

In this case, the usefulness of the quantum entanglement principle can be illustrated on the example of solving the *Quantum Monty Hall problem* [14]. In the classical case (without quantum effects), co-worker **A** knows the true idea, which will be of interest to the chief, but co-worker **B** does not. Co-worker **B** selects one of the three ideas. Next, co-worker **A** calls a different idea, showing that he is not sure about the true idea. Then co-worker **B** has the option of keeping his previous choice or changing it. The optimal strategy for **B** is to change his choice, thereby doubling his chance to recognize the true idea, from 1/3 to 2/3. In this situation, the null operation does provide information about the problem.

A quantum version of the Monty Hall decision problem is as follows. There exist three different ideas, $|0\rangle$, $|1\rangle$, and $|2\rangle$, and only one of them true (and known to co-worker **A**). Co-worker **A** selects the superposition of the three ideas, trying to hide the true idea in an accidental way. **A** will be called swindler. Next, co-worker **B** selects one of the three ideas. It makes the game fair due to introducing an additional idea that is entangled with the true one. This entangled idea can be introduced using the criteria in accordance with the goal of the meeting announced by chief **C**. It helps **A** to make a quantum measurement of the ideas based on these criteria.

The state of the quantum system, an interpretation of this decision problem, is written as $|\psi\rangle = |o, b, a\rangle$, where a and b denote the choices of co-workers **A** and **B**, respectively, and o is the idea that chief **C** wants to hear. As was proved in [14], if one takes the initial state with the maximum entanglement between the choices of **A** and **B**, then **B** has access to a quantum strategy, whereas **A** does not. For example, assume that chief **C**, who invited co-workers to generate a true idea, also announced the exact goal and criteria. In this case, **B** can recognize the true idea and win all the time.

Without entanglement, the quantum game confirms the expectations by offering only classical mixed strategy; **B** is able to deceive **A** with a high level of probability. Moreover, co-worker **B** wins in 2/3 of cases by changing his choice when both co-workers have access to quantum strategies (both know the real goal and criteria), and the maximum entanglement of the initial states yields the same payoff as in the classical game.

Thus, quantum semantics appear to be useful for aggregating the ill-defined component of group members in collective decision-making, i.e., for describing the unreasonable and uncaused decisions of participants. An important innovation of this approach consists in the cognitive semantic interpretation of messages, which are generated in virtual conversations. The special quantum operators generate new knowledge and assist in penetrating into recognized objects. This looks like a procedure for identifying the internal genotype structure of an organism by its external image of phenotype.

At the same time, numerous researches into the issue of quantum computation are based on the superposition of binary data, i.e., have a digital foundation. For example, in the papers [15, 16], the concept of timeless quantum networks was considered. The effects of entanglement, teleportation, quantum cryptography, dense coding, etc., were studied. The syntax and semantics for a new Q-bit were presented using category theory. The mechanism of graphical calculations was developed to ensure the correctness of generalized protocols. However, the approaches proposed therein do not change the fundamental binary principle of data representation. The quantum superposition only interpolates them with some accuracy.

Thus, the main problem is that modern quantum calculations rest on the binary Q-bit model, which tears the noosphere and biosphere nature of consciousness and thinking into small pieces. Perhaps, the representations of these phenomena by Q-bits are very scarce and need to be augmented with different tools, more continuous and smooth. To proceed, consider in detail the continuity of the space of consciousness from the viewpoint of optics.

References

1. Aerts, D., Czachor, M.: Quantum Aspects of Semantic Analysis and Symbolic Artificial Intelligence. J. Phys. A.: Math. Gen. **37**(12), L123–L132 (2004)
2. Atmanspacher, H.: Quantum Approaches to Brain and Mind. An Overview with Representative Examples. In: The Blackwell Companion to Consciousness, Schneider, S. and Velmans, M. (Eds). John Wiley & Sons Ltd., Hoboken, NJ., pp. 298–313 (2017). https://doi.org/10.1002/9781119132363.ch21
3. Ivancova, O.V., Korenkov, V.V., Ulyanov S.V.: Quantum Software Engineering. Quantum Supremacy Modelling. Part I: Design IT and Information Analysis of Quantum Algorithms: Educational and Methodical Textbook, Dubna, Joint Institute for Nuclear Researches, INESYS (EFCO Group), Moscow, KURS, 328 p. (2020)
4. Ivancova, O.V., Korenkov V.V., Ulyanov S.V.: Quantum Software Engineering. Quantum Supremacy Modelling. Part II: Quantum Search Algorithms Simulator—Computational Intelligence Toolkit: Educational and Methodical Textbook, Dubna, Joint Institute for Nuclear Researches, INESYS (EFCO Group), Moscow, KURS, 344 p. (2020)
5. Raikov, A.: Post-non-Classical Artificial Intelligence and its Pioneer Practical Application. Part of Special Issue. 19th IFAC Conference on Technology, Culture, and International Stability (TECIS 2019), Sozopol, Bulgaria, IFAC-PapersOnLine. vol. 52(25), pp. 343–348 (2019). https://doi.org/10.1016/j.ifacol.2019.12.547
6. Ulyanov, S.V.: Quantum Fast Algorithm Computational Intelligence PT I: SW/HW smart toolkit. Artif. Intell. Adv. **1**(1) (2019). https://doi.org/10.30564/aia.v1i1.619
7. Raikov, A.: Strategic Analysis of the Long-Term Future Needs of Educational Services. In: Proceedings of 3rd World Conference on Smart Trends in Systems, Security and Sustainability (WorldS4 2019). Roding Building, London Metropolitan University, London, IEEE, pp. 29–36, (2019). https://doi.org/10.1109/WorldS4.2019.8903983
8. Krpan, D., Kullich, W.: Nuclear Magnetic Resonance Therapy (MBST) in the Treatment of Osteoporosis. Case Report Study. Clin. Cases Miner. Bone Metab. May-Aug; vol. 14(2), pp. 235–238 (2017). https://doi.org/10.11138/ccmbm/2017.14.1.235
9. Dalela, A.: Quantum Meaning: A Semantic Interpretation of Quantum Theory. Shabda Press (2012)
10. Raikov, A.N.: Holistic Discourse in the Network Cognitive Modeling. J. Math. Fundam. Sci. **3**, 519–530 (2013)
11. Zipf, G.K.: Human Behavior and the Principle of Least Effort. Addison-Wesley. Online text, https://archive.org/details/in.ernet.dli.2015.90211/page/n327/mode/2up (1949). Accessed December 7, 2020
12. Eisert, J., Wilkens, M., Lewenstain, M.: Quantum Games and Quantum Strategies. Phys. Rev. Lett. **83**(15), 3077–3088 (1999)
13. Litvintseva, L.V., Ulyanov, S.V.: Intelligent Control System. I. Quantum Computing and Self-organization Algorithm. J. Comput. Syst. Sci. Intern. **48**(6), 946–984 (2009)

14. Flitney, A.P., Abbott, D.: Quantum Version of the Monty Hall Problem. Phys. Rev. A **65**(6), (06318)1–4 (2002)
15. Borrill, P.: Stanford Seminar: the Time-less Datacenter https://www.youtube.com/watch?v= IPTlTmH-YvQ (2016). Accessed December 7, 2020
16. Reutter, D., Vicary, J.: A Classical Groupoid Model for Quantum Networks. Logical Methods Comput. Sci. **15**(1), 32:1–32:28. https://arxiv.org/pdf/1707.00966.pdf (2019). Accessed December 7, 2020

Chapter 11
Optical Semantics

A discrete representation of data can be replaced by a continuous (analog) one using optical tools. Such a substitution is not absolute since optical transformations have a corpuscular-wave nature. The corpuscular properties of light can be described by the quantization procedure; see Chap. 21. As is well known, a quantum particle of light is a photon with zero rest mass and spin 1, carrying no electric charge. The quantum–mechanical aspect of an electromagnetic field of light is characterized by the wave function, which remains the same under an interchange of two photons: they are bosons and obey *the Bose–Einstein statistics*. At the same time, the replacement of the binary form of primary data in a digital computer by the corpuscular-wave one has a fundamental difference. Note that a quantum bit also has a binary basis.

The development of an optical approach is dictated by the following features: modern supercomputers can solve a very complex problem for months, but sometimes problem-solving is required to take no more than a few seconds. A quantum computer (QC) will most likely solve a complex problem faster than a supercomputer, by several times but not by several orders of magnitude. The restrictions may follow from the reliance on the digital representation of data or on the superposition of bits (Q-bits). The semantics of AI models in modern quantum computers is represented by the superposition of digital records, which is denotative semantics. Consequently, cognitive semantics cannot be fully embraced by QCs.

For accelerating computations by several orders of magnitude, one perhaps needs to reject the binary representation of data and operate continuous data with optical transformation tools. This can be provided by an optical computer (OC), which is intended to process continuous signals with an infinite natural spectrum. In this case, the time of solving a complex problem may become comparable with the time of light propagation to several meters.

The fundamental difference between OCs (on the one part) and classical digital computers and quantum computers (on the other) is the rejection of the discrete representation of a signal. This can be illustrated, for example, using the *Sampling Theorem*, attributed to Kotelnikov, Nyquist, and Shannon. The question could be

A. Raikov, *Cognitive Semantics of Artificial Intelligence: A New Perspective*,
SpringerBriefs in Computational Intelligence,
https://doi.org/10.1007/978-981-33-6750-0_11

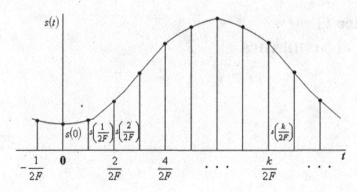

Fig. 11.1 Signal and its samples

formulated in a broader way: at what level of completeness and holistic can the problem be represented by means of digital signals?

The source of information can generate signals: first, in the digital forms like symbols, bits, bytes, words, images, schemes, and sentences; and second, in a phenomenological form like meanings, emotions, intentions, unconscious, etc. The latter form belongs to the aspects of cognitive semantics. The digital forms of a signal can obviously be transmitted via traditional data channels and computer transformations with a specified accuracy, but the phenomenological ones cannot.

Perhaps, this comparison is not correct, as phenomenological meanings and digital words lie in different spaces. The measurement of thoughts cannot be reduced to the direct processing of a digital signal: thoughts are essentially more complex phenomena than verbal and matter entities. For example, a linguistic sentence describes an object. But its phenomenological meaning can only be explicated through unformalized paradoxes, intentions, free will, etc.

According to the Sampling Theorem, a signal $s(t)$ *with a limited spectrum* can be accurately reconstructed from its samples $s(k\Delta t)$ taken at intervals $\Delta t = 1/(2F)$, where F is the upper frequency of the signal spectrum and k is the integer; see Fig. 11.1. This signal can be represented by a sequence of samples at the discrete points $k/2F$.

The spectrum of a continuous signal transmitted via communication channels can be trimmed when some signal distortion is allowed. Only a signal with a limited spectrum can be transmitted accurately. However, thoughts, feelings, and meanings are phenomena with an infinite spectrum, and they cannot be encoded by digital signals very accurately. One of the approaches to remove these digital paradigm-induced limitations for the transmission of phenomenological information is its complement by means of continuous or analog signals.

For example, when implementing visualization tools, a virtual context with 3D models is formed. As is well known, visualization assists in problem-solving. Visual analytics can be defined as an integrated approach that employs both visualization

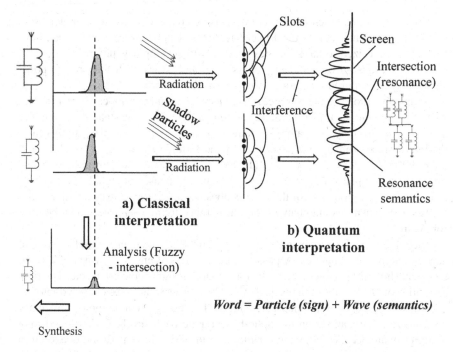

a) Classical interpretation

Analysis (Fuzzy - intersection)

b) Quantum interpretation

Word = Particle (sign) + Wave (semantics)

Synthesis

Fig. 11.2 Analog and discrete processes compared with one another

and computer processing of data. But in the case of transmitting phenomenological information, a processed image has to be analog and equipped with cognitive semantics.

An algorithm for the visual processing and analysis of images can be rigorously defined as a transformation $F: S \to I$, where S is some data for analysis and I is an idea produced by this analysis. Here S means data in a broad sense, including both of these processes. A possible way to emulate these processes is creating a "resonant semantic tool" for the cognitive semantic interpretation of symbols (words). In Fig. 11.2, the analog and discrete processes in this tool are compared with one another. This figure demonstrates two different versions of semantics interpretations. Each word can be transmitted simultaneously as a particle and as a wave. Two different words in a computer do not literally coincide, but they can be intersecting by meaning. If the meanings of such words are encoded in the form of waves, then their interference generates a new signal with the corresponding resonance curve; see Fig. 11.2a.

The quantum semantic approach makes the interpretation picture more complete. Each particle (boson) will interact with an infinite number of the same (albeit, hidden or shadow) particles; like in the well-known two-slit quantum experiment (also, see Fig. 10.1), the complex interference pattern of meanings will be created on the screen (Fig. 11.2b). This effect cannot be reproduced with absolute accuracy by a digital computer: in this case, the matter concerns only an approximate simulation model for the solution of the Maxwell and Schrödinger equations.

This can be done much more accurately using an analog-optical computer. Note that such a computer needs novel materials for optical transformation, which have not been synthesized so far. Also, it is necessary to develop new physical and mathematical methods for optical data processing. Finally, various distortions of optical signals, e.g., aberrations, cannot be completely eliminated.

At the same time, the interference of signals in quantum physics can be controlled. As is well known, the transition of a quantum particle from one state $|x>$ to another $|y>$ can be realized by two paths simultaneously: $\langle y|x \rangle = \langle y|x \rangle_1 + \langle y|x \rangle_2$. In this case, the interference probability P_{xy} is given by the formula $P_{xy} = P_{xy}(1) + P_{xy}(2) + 2\,Re\,[e^{i(\theta-\omega)}\langle y|x \rangle_1 \langle y|x \rangle_2]$, where $P_{xy}(1)$ and $P_{xy}(2)$ denote the probabilities of the first and second paths, respectively, θ and ω are the phases of signals. A variation in the phases or amplitudes of the signals does not affect the probability of the first and second transitions but changes the interference term. Therefore, it can be used for control.

These ideas can be implemented for designing OCs, which will partially take into account cognitive semantics. A possible architecture of an OC consists of a laser, rewritable holographic memory, deflectors, coherent signal converters, etc. The time resolution of OC memory can reach 10^{-15} s. The Fourier transforms can be adopted as a methodological basis for optical calculations. The theory of autonomous elementary computational operations on the optic-holographic basis has been developed long ago. For example, the following operations can be implemented: the convolution and differentiation of a function, the Fresnel transform, the recovery of a function from the spectral density of its sum, and the subtraction of functions. Such operations allow for replacing binary data with analog ones. But the materials known in the Earth for creating the multidimensional rewritable holographic memory of an OC, do not meet the following necessary requirements:

- Fourier images-based multilayer data recording, with the location of about 100 different images at one holographic point via multiple overlays of the objects' images and references beams;
- fast (femtosecond-range) overwriting of each image, with an appropriately changed direction of the reference beam;
- synchronous propagation of the reference and subject's beams for recording and reading images;
- the sustainability of holograms during reading, which can be achieved by the stability of switching polarization;
- saving the focus settings for optical elements in the case of physical disturbances (shaking, bumps, etc.), and others.

The creation of OC requires new optical mathematics. This statement can be explained by the following example. A 3D map of a city contains many layers. Some of the layers reflect the density of residents in different areas of the city. The problem is to find an optimal arrangement of kindergartens in "access zones" under given norms.

This problem admits of several solutions. For example, one solution involves the convolution of functions based on optical computing:

$$f(x) * s(x) \equiv \int_{-\infty}^{\infty} f(\xi)s(x - \xi)d\xi, \qquad (11.1)$$

where $f(x)$ and $s(x)$ are the density of residents and the norm specifying the location of kindergartens in access zones, respectively, ξ is real number. The convolution will determine the "similarity" of these functions, indicating an approximate arrangement of the kindergartens. For this, the Borel theorem on the convolution [1] is applicable:

$$\begin{aligned}\mathcal{F}[f(x)]\mathcal{F}[s(x)] &= \mathcal{F}[f(x) * s(x)], \\ \mathcal{F}[f(x)s(x)] &= \mathcal{F}[f(x)] * \mathcal{F}[s(x)], \end{aligned} \qquad (11.2)$$

where $\mathcal{F}[\varphi(x)]$ denotes the Fourier transform of a function $\varphi(x)$:

$$\mathcal{F}[\varphi(x)] = \int_{-\infty}^{\infty} \varphi(x) \exp(-2\pi i \xi x)dx, \qquad (11.3)$$

For performing the convolution of two functions, one has to calculate the Fourier transform of these functions, obtain $G(\omega)$ by using (11.2), and then calculate the inverse Fourier transform:

$$g(x) = \mathcal{F}^{-1}[G(\xi)] = \int_{-\infty}^{\infty} G(\xi)\exp(2\pi i \xi x)d\xi. \qquad (11.4)$$

An OC is expected to integrate the architectures of a supercomputer (digital calculations) and a traditional quantum computer with an optical processor for image-analog processing. As an example, a possible scheme of an optical processor for solving the optimal arrangement problem of objects on a map is shown in Fig. 11.3.

There is another application: this optical basis can be used to design a neural processor. The core of this device is an optical neural network with a matrix structure of optical elements (cells) that emulate the behavior of a natural neuron. For such an optical element, the output brightness is proportional to the brightness of the incoming optical signals, and this brightness is distributed over many connections with other optical cells. In the future, such a cell can be built on the rewritable holographic memory described above, with the possibility of its multifunctional modulation.

To construct axons, it is necessary to develop ultra-thin optical fibers and holographic lenses, whose light conductivity can be changed by external influence. A neural network is trained by configuring this element.

The device input can be a "laser array" connected to holographic transducers. This can be a source of coherent radiation, a beam expander, polarizers, deflectors, telescopic units, semitransparent mirrors (used to form reference and object beams), lenses, etc. Experiments with light-emitting diodes are possible.

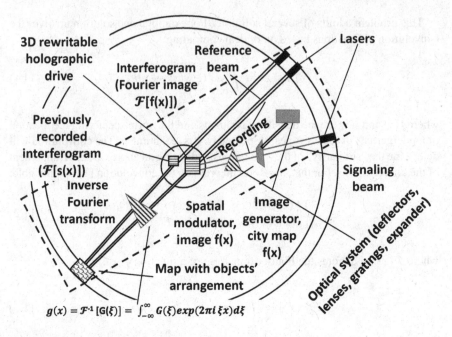

Fig. 11.3 Scheme of optical processor

An optical neural network can be trained in a traditional way using digital super-computers. At the same time, apparently, an algorithm can be developed for training a neural network almost instantly.

The device must have an analog-to-digital interface. A special unit has to be developed to control the device. The device can be controlled using a quantum computer or a classical supercomputer.

These optical computing tools will help to overcome the existing restrictions on the speed of calculations. In particular, they will be useful in complex (inverse, multidimensional) computing, real-time image processing, hydrodynamic optimiza-tion, satellite sounding data processing, object recognition, meteorological data processing, aircraft and spacecraft control, the design of autonomous systems and self-governing groups of objects, capable of independent decision-making and self-programming in real time, the analysis of large volumes of non-structured data, the identification of implicit dependencies, including in disaster medicine, and other areas of application.

Reference

1. Stromberg, K.: A Note on the Convolution of Regular Measures. Math. Scand **7**(2), 347–352 (1960)

Chapter 12
Wave of Consciousness

Human consciousness can be compared with a physical field. Apparently, the low-frequency signals that the brain sends outwards are communication signals. However, these signals rather resemble the speech signals of a human who verbally transmits his or her thoughts to another human. Thought itself is a much more complex phenomenon.

One can try to express this complexity through the biochemical, wave, acoustic, and quantum-relativistic nature of consciousness. Then the wave aspect can be considered, in addition to the well-known low-frequency signaling device, as well as acoustic, electromagnetic, and quantum-relativistic correlators. Under the impact of an external wave field, the human body and brain can interact with this field, being a kind of a resonant receiver, a translator, and a generator.

Now let us discuss the fundamental principles of describing fields and the wave nature of signal propagation. Here the key tools are the d'Alembert and Helmholtz equations, the Green function, and the formalisms of electrodynamics, both classical and quantum. In this case, "thought events" can be represented in some infinite-dimensional and unbounded space (e.g., a vector space), in which some points can be emitting a signal (wave, quantum, sound), a source point x', and the other can be the result of emission, a point x. Thought events occur in time t. Impacts can be applied in forward and backward time, i.e., direct and inverse problems can be solved.

In the general case, the D'Alembert equation for scalar waves [1] can be the starting point for solving direct and inverse problems in a wave medium:

$$\nabla^2 u(x, t) - \frac{1}{c(x)^2} \frac{\partial^2}{\partial t^2} u(x, t) = F(x, t), \qquad (12.1)$$

where ∇^2 denotes the Laplace operator (the sum of the second partial derivatives of a function with respect to spatial coordinates); $u(x,t)$ is the requisite function of spatial

© The Author(s), under exclusive license to Springer Nature Singapore Pte Ltd. 2021
A. Raikov, *Cognitive Semantics of Artificial Intelligence: A New Perspective*,
SpringerBriefs in Computational Intelligence,
https://doi.org/10.1007/978-981-33-6750-0_12

coordinates and time; c gives phase velocity; x is the radius vector of an arbitrary point in the coordinate space; finally, $F(x,t)$ is a time-varying radiation source. This equation holds in both real and complex representations.

Equation (12.1) shows the possibility of expanding the field of semantic interpretation of thought processes. It allows for an inhomogeneous medium, i.e., the dependence of the signal propagation velocity on different factors, including the coordinates and the impact of the observer (participants). The behavior of the field can be quantized, analog (continuous), and discrete.

The scalar representation $u(x,t)$ has different physical meanings. For example, it can describe the magnitude of the electromagnetic field (both in its classical and quantum manifestations), a gravitational wave, sound pressure, and other oscillatory processes in a continuous medium, including a cosmic vacuum. Such processes can affect thinking.

As is well known, the solution of Eq. (12.1) can be written as the sum of traveling waves or, when using the Fourier expansion, as an infinite linear combination of ordinary harmonic functions. In this case, the spectrum of the function $u(x,\omega)$, where ω is the frequency, will have the form:

$$u(x, \omega) = \int_{-\infty}^{\infty} u(x, t) \exp(-i\omega t) dt. \tag{12.2}$$

Then the signal is restored from the spectrum:

$$u(x, t) = \frac{1}{2\pi} \int_{-\infty}^{\infty} u(x, \omega) \exp(-i\omega t) d\omega. \tag{12.3}$$

Note that using formula (12.3), the field sources can be written as the Fourier integral with respect to their harmonic components:

$$F(x, t) = \frac{1}{2\pi} \int_{-\infty}^{\infty} F(x, \omega) \exp(-i\omega t) d\omega. \tag{12.4}$$

According to (12.2) and (12.3), a continuous analog signal $u_a(x, t)$ takes the form of a complex number and can be represented by adding to the original function u (x, t) the imaginary part involving the Hilbert transform $u_H(x, t)$:

$$u_a(x, t) = u(x, t) - i u_H(x, t). \tag{12.5}$$

Obviously, the spectrum of the continuous signal is nonzero only for $\omega > 0$. Thus, the field $u(x, \omega)$ characterized by the wave number $k(x) = \omega/c(x)$, is created by the source $F(x, t)$, and Eq. (12.1) turns into the Helmholtz equation:

$$\Delta u(x, \omega) + k^2(x) u(x, \omega) = F(x, \omega). \tag{12.6}$$

When solving this equation, one uses the Green function $G(x, x', \omega)$, representing a field created at a point x by a source located at a point x'. This function corresponds to a wave diverging at infinity.

The physical meaning of the Green function is as follows. With a field interpretation of thought processes, this function reflects its behavior at two points in the space of thought events (the source point x' and another point x) only at a single frequency ω. Therefore, note that to solve problems for thought processes, this function has to be expanded to an infinite spectrum of frequencies.

The Green function can have lag or advance, depending on the convergence or divergence of the wave at infinity. With the Green function, the solution of Eq. (12.6) can be written as

$$u(x, \omega) = \int_{-\infty}^{\infty} G(x, x', \omega) F(x', \omega) dx'. \tag{12.7}$$

Thus, the spatial distribution function of radiation sources, $F(\mathbf{r}, t) = F(\mathbf{r}) exp(-i\omega t)$, can be restored from the externally observed field $u(y)$ by solving an inverse problem with respect to $F(\mathbf{r})$ of the form:

$$u(y) = \int_{\mathbb{R}} G(y, r') F(r) dr, \, y \in \mathbb{R}. \tag{12.8}$$

Here $\mathbb{R}$ is a domain of unknown radiation sources (e.g., in distant cosmos), y is an observable domain separated from the radiation domain $\mathbb{R}$; $G(y,r)$ is the Green function. This problem is incorrect: the non-uniqueness of its solution is determined by the possible presence of many unknown sources. At the same time, imposing additional conditions on radiation, e.g., the Sommerfeld conditions, will ensure the unique solution [2].

In applications, the source and receiver of radiation can be located on one or both sides of the physical body. The one side is the human body (or the human brain only), and the other is the external environment, which can extend far enough.

The probability of fixing a certain configuration of sources and receivers of signals (particles) in the limited space of the human brain is negligible. In view of the number of subatomic particles in the human brain, this probability is expected not to exceed 10^{-25} without external pressure (i.e., assuming the most chaotic state of a multitude of particles). Note that the behavior of these particles can be stochastic. However, physical regularities and gradients, genetic and evolutionary mechanisms, most likely, predetermine some non-random systematization of these particles at certain levels. Then the probability under consideration increases by several orders of magnitude, i.e., by a value commensurate with the number of neurons and their interconnections in the human brain.

The formalisms discussed above touch upon the initial aspects of the wave representation of thought processes. They have been presented here in order to demonstrate the following: when fixing the presence of a radiation source or the result of radiation, each point of the thought space (distributed between the human body and its environment) identifies an event reflecting a phenomenon of definite nature. This can be, for example, the state of a quantum particle, whose wave and corpuscular behavior obeys nonlocal laws.

References

1. Pike, R., Sabatier, P.C. (Eds.): Scattering: Scattering and Inverse Scattering in Pure and Applied Science, p. 1831. Academic Press, San Diego, San Francisco, New York, Boston, London, Sydney, Tokyo (2002)
2. Colton, D., Kress, R.: Inverse Acoustic and Electromagnetic Scattering Theory. Applied Mathematical Sciences, vol. 93. Springer-Verlag, New York, XIV. https://doi.org/10.1007/978-1-4614-4942-3 (2013)

Chapter 13
Entanglement of Thought

The methods of quantum semantics seem to be promising for the construction of cognitive interpretations of AI models created in the course of thinking. Separate studies of this issue can be found in [1–3]. Of particular interest are two effects: the collapse of quantum states and the entanglement of quantum particles.

The collapse of quantum states occurs during measurement when an observer is involved in a quantum process. At this time instant, one of many admissible states of a quantum particle becomes "frozen." This effect resembles the "Eureka" effect [4], when a human suddenly finds a solution of a problem after long reflections due to an unexpected external impulse; also, see Chap. 7.

The entanglement of quantum particles reflects an instantaneous relationship between the states of particles located at large distances. This is also called the nonlocality effect. This effect of quantum states has different explanations generated by the well-known thought experiment (1935), called the Einstein–Podolsky–Rosen (EPR) paradox. Here is an interpretation of this phenomenon [5]. Consider two quantum particles, one of which is located in a human neuron, and the other anywhere in the Universe. Let the first and second particles take two different states, $|u_1\rangle$, $|u_2\rangle$, and $|v_1\rangle$, $|v_2\rangle$, respectively. Thus, a quantum system consisting of these particles can be represented as the superposition of states:

$$(\alpha_1|u_1\rangle + \alpha_2|u_2\rangle) \otimes (\beta_1|v_1\rangle + \beta_2|v_2\rangle) =$$
$$\alpha_1\beta_1|u_1\rangle \otimes |v_1\rangle + \alpha_1\beta_2|u_1\rangle \otimes |v_2\rangle + \alpha_2\beta_1|u_2\rangle \otimes |v_1\rangle \quad (13.1)$$
$$+\alpha_2\beta_2|u_2\rangle \otimes |v_2\rangle,$$

where $\alpha_1, \alpha_2, \beta_1$, and β_2 are some coefficients from the set of real numbers; $\otimes$ denotes the tensor product. At the same time, the state of this quantum system may have an alternative description—the sum of the superpositions of the states:

$$|u_1\rangle \otimes |v_1\rangle + |u_2\rangle \otimes |v_2\rangle. \quad (13.2)$$

© The Author(s), under exclusive license to Springer Nature Singapore Pte Ltd. 2021
A. Raikov, *Cognitive Semantics of Artificial Intelligence: A New Perspective*,
SpringerBriefs in Computational Intelligence,
https://doi.org/10.1007/978-981-33-6750-0_13

Moreover, the products in (13.2) are non-commutative. Clearly, the state (13.2) is non-factorizable, i.e., cannot be reduced to the product of states of two particles. Indeed, in the more general Eq. (13.1), it is impossible to select appropriate values of the coefficients α_1, α_2, β_1, and β_2. According to (13.2), different observers, each observing only one of the particles, cannot control its behavior. These states are "fixed" at a definite time instant. This instant comes when one of the observers "sees" the state of its particle.

Well, consider two observers (humans, usually called Bob and Alice), each observing his or her own particle. According to formula (13.2), at the time instant when one of them (say, Bob) fixes the state of his particle, a quantum collapse occurs as follows: at the same instant, the second observer (Alice) is doomed to see only one of the states of her particle. It may be either the state that is fixed in (13.2) before the "+" sign or the state after this sign. Note that no information is transmitted from one observer to another since this would, in principle, contradict the theory of relativity.

However, the given explanation may not seem completely correct. For example, the time instant of observations is considered the same. Even if one admits the synchronization of two observations, which can now be done with an accuracy of up to 10^{-18} s (the accuracy of an atomic clock), it turns out that the collapse occurs with some accuracy in time, and not with absolute accuracy. Assuming such synchronization, one also admits the idea of instantaneous information transfer (or with speed significantly exceeding the velocity of light). Although, this aspect of considering the effect can be omitted: prior to the time instant of observation, the quantum (thought) system was in the state of superposition; it enters the state of collapse at the time instant when Bob observes it; the second observer, Alice, can do this a little bit later. In other words, this "subtle" temporal aspect of an instantaneous event might be neglected, and then the canons of the theory of relativity would not be violated. Meanwhile, when measuring the properties of a quantum particle, the idea that the observed state actually existed even before the measurement may turn out to be incorrect [6]. The development of this issue should be traced back. In particular, one should consider a formalization of this effect for the multidimensional case, suggested by Bell in 1964. Recently, Bell tests and experiments demonstrated strong disagreement with local realism, in which the properties of the physical world are independent of our observation of them, and no signal travels faster than light [7].

Concerning the applicability of quantum physics to the analysis of the processes of consciousness and thinking, note that some fundamental laws of quantum physics have significant limitations. For example, the Schrödinger equation cannot:

- explain spontaneous emission since the wave function is the exact time-dependent solution;
- describe the measurement process since the equation is linear, deterministic, and reversible in time;
- represent the processes of mutual transformations of elementary particles.

This book merely indicates the place of such questions, showing possible limitations of any well-known regularities for a holistic description of human consciousness and thinking. In particular, as has been illustrated by the example with quantum nonlocality, thought's event may indirectly and unreasonably depend on the distant environment.

References

1. Dalela, A.: Quantum Meaning: A Semantic Interpretation of Quantum Theory. Shabda Press (2012)
2. Raikov, A.: Convergent Networked Decision-Making Using Group Insights. Complex Intell. Syst. **1**, 57–68 (2015). https://doi.org/10.1007/s40747-016–0005-9
3. Raikov, A.: Cognitive Modelling Quality Rising by Applying Quantum and Optical Semantic Approaches. In: Proceeding of 18th IFAC Conference on Technology, Culture, and International Stability. IFAC-PapersOnLine, vol. 51(30), pp. 492–497. Baku, Azerbaijan (2018) https://doi.org/10.1016/j.ifacol.2018.11.309
4. Perkins, D.: The Eureka Effect. The Art and Logic of Breakthrough Thinking. Norton, New York, London (2001)
5. Zwiebach, B.: Entanglement. https://ocw.mit.edu/courses/physics/8-04-quantum-physics-i-spring-2016/video-lectures/part-1/entanglement/ (2016) Accessed December 7, 2020
6. Scheidl, T., et al.: Violation of local realism with freedom of choice. PNAS **107**(46), 19708–19713 (2010). https://doi.org/10.1073/pnas.1002780107
7. The BIG Bell Test Collaboration., Abellán, C., Acín, A. et al. Challenging Local Realism with Human Choices. Nature 557, 212–216 (2018). https://doi.org/10.1038/s41586-018-0085-3

Chapter 14
Thought Event

The central element in the cognitive space of individual or group thoughts can be based on the concept of an "event." It is a spiritual-physical phenomenon in some mental space with limited scope and time. Otherwise, a cognitive thought could not be exactly identified and then verbally expressed with a high level of accuracy. A thought event is limited, but not a δ-function and not everything. It has some imprecise or fuzzy boundaries because any quality has a measure in which it exists.

This event, with its coordinates and temporal aspects, depends on an observer who wants to (but cannot) represent it in a direct and logical way. The reasons consist in the following: as soon as a verbal description is assigned to a thought, the latter remains outside this description, as a physical and energetic phenomenon. A thought is in a space (field), but its description is on the table (symbols and words). Special mathematical abstractions are required for separating an event from an observer, for making the former independent of the latter.

Each event can be associated with a point in the space representing some mathematical object, separable from other objects. This separability can be described in a nonmetric topological space. The point's coordinates are not important, as they only provide some means for its representation. There may exist different means of this kind, each without an obvious advantage over the others.

Time and place are the components of the space design. Suppose that the Minkowski space is relevant here. It is a four-dimensional manifold where the interval between any two events can be preserved under various spatial transformations. What is important, this space can be adapted for electromagnetic applications and take into account the postulates of the theory of relativity and quantum physics.

For example, using the tools of quantum physics to represent cognitive semantics will require increasing the dimension of the space to infinity (a Hilbert space can be used). For considering the nonlocal behavior of subatomic components in the organic components of thinking (neurons, brain, body), one should take into account the aspects of the theory of relativity. For more complete and detailed coverage of the wave, quantum, and relativistic aspects in the cognitive semantic representation

A. Raikov, *Cognitive Semantics of Artificial Intelligence: A New Perspective*,
SpringerBriefs in Computational Intelligence,
https://doi.org/10.1007/978-981-33-6750-0_14

of thought processes, a comprehensive study of these processes using group theory can be performed.

In this case, the elements of spatial sets will be various entities: events, ideal and virtual objects, complex numbers, and vectors with their scalar products. Each event will be assigned a point in the space, treated as a mathematical object differing from others. Thus, the Fourier transforms and also the transformations over the elements of the 4-dimensional Minkowski space of events can be covered: the points in this space can have a qualitative (non-numerical) character, and the coordinates of these elements can be a way for specifying these points. Following the conventional approach, such a space will be denoted by $\mathcal{M}$, and points in this space by P, Q, etc.

The group theory approach seems useful for describing this situation. This approach allows placing the symbolic model with its denotative (formalizable) semantics and cognitive (non-formalizable) semantics into a single space and using unified operators to transform events of various natures. In this case, events can be attributed (not completely) to the phenomena described by the laws of logic and also by the laws of quantum electrodynamics and the theory of relativity. If cognitive semantics admits a quantum interpretation with its representation in a Hilbert space, as well as a relativistic interpretation with its representation in terms of transformations of smooth manifolds (e.g., superstrings [1]), then group theory seems to be fruitful for the indirect representation of non-formalizable cognitive semantics.

At the same time, the adequacy of the group theory-based approach to the interpretation of cognitive semantics is subject to verification, especially with regard to its applicability for a joint study of, e.g., events reflecting the behavior of hypothetical superstrings (quantum physics) and cosmic strings (astrophysics).

Other possibly useful tools include transformations in the Lorentz group, in which a certain starting point is fixed, and the intervals between events (points) are preserved, or transformations in the Poincaré group, where fixed points are not required. Then a fixed point will interpret the binding of a particular thought process to a certain subject or group of people, and the same intervals between events will determine the concentration of the thought process on a certain task.

The Lorentz group is a convenient mathematical base, as the theory of relativity or gauge theory. The latter is a type of quantum field theory in which the Lagrangian, describing the states of a quantum field, has an infinite number of degrees of freedom, being invariant under local transformations from the Lie groups. A Lie group is a group in which elements are organized continuously and smoothly, as opposed to discrete groups, where the elements are separated. This makes Lie groups smooth manifolds [2].

Within the notion of compact groups, particles are considered as the members of the multiplets of the groups' representations, but compactness is a necessary condition for correct inverse problem-solving [3]; also, see Chap. 18. A multiplet represents a mathematical structure of a Lie group acting as an operator on a real or complex vector space. For example, group theory helps to describe string theory—a new paradigm trying to resolve the problem of quantum gravity. The basic element of string theory is an extended "one-dimensional" object of a diameter less than

10^{-20}cm; it can be open or closed, and its vibration induces electromagnetic fields [1, 4].

Thus, group theory can be a fairly universal approach to construct the cognitive semantics of AI models that covers their features at the micro and macro levels.

Definition of a group. A non-empty set G with one binary operation is called a group if:

(1)   The operation in G is associative.
(2)   The reverse operation in G is feasible.

Note that this definition can be employed to substantiate, in a rather simple way, one important property of the cognitive transformations under study. This will be done proceeding from the fact that a group can be finite or infinite [5]. For a finite group, the solution of the direct and inverse equations may be non-unique; for an infinite group, it may be unique. For a finite group, condition (2) can be reduced to the uniqueness of the solution of the direct and inverse equations. Well, let the set G be with one operation and unique solutions of these equations (if they exist). Consider two different elements from G. Multiply the first of these elements on the right by an element x from G, and let x be running through all elements from G. As a result, a finite number of different elements from G is obtained. There exists an element x_0 such that its product with one of the selected elements will give a solution of the direct equation. The existence of a solution of the second (inverse) equation can be established by analogy.

For the infinite space of cognitive semantic interpretation of the AI model (thoughts, feelings, free will, etc.), such a weakening of condition (2) becomes inadmissible. If the solution of the direct problem exists and is unique, then the solution of the inverse problem can have different values (local solutions), each possibly initiated by even a very small variation in the initial data. These considerations lead to the possibly incorrect solution of the inverse problem in the infinite case (a possible occurrence of many solutions).

In the case of the infinite space, it may be fruitful to use category theory [6] and to justify necessary conditions for the correct solution of inverse problems in a topological space. It makes any collective strategic decision-making process purposeful and sustainable; see [7] and Chap. 18.

For the problem in the infinite space, allowing for transformations with a single fixed point (the Lorentz group), it is therefore convenient to form some convolutional kernel, e.g., in the form of a subset of G generating G. In quantum physics, it is called a gauge; in astrophysics, it is associated with a convolutional filter (e.g., Haar or Gabor wavelets [8]). In AI, a whole class of transformations is also called convolutional; for example, convolutional neural networks.

The construction of such a kernel is partly done to decrease the amount of computations. On the one hand, it helps to process a large-scale event (phenomenon) compactly; on the other, it irrevocably reduces this event, making the latter restorable only with a certain accuracy. While creating the convenience of computer representation and visibility of an event, this filtering or convolution can however distort its natural spectrum. Such an approach may be unacceptable for solving complex problems based on the analysis of big data, e.g., the ones obtained by an LHC or radio telescope.

References

1. Sazhina, O.S., et al.: Optical Analysis of a CMB Cosmic String Candidate. Adv. Access Publ MNRAS **485**, 1876–1885 (2019). https://doi.org/10.1093/mnras/stz527
2. Lee, J. M.: Introduction to Smooth Manifolds. Springer-Verlag, New York (2000). https://doi.org/https://doi.org/10.1007/978-1-4419-9982-5
3. Ivanov, V.K.: Incorrect Problems in Topological Spaces. Siberian Math. J. 10, 785–791 (1969). Novosibirsk, Russia
4. Yau, S.-T., Nadis, S.: The Shape of Inner Space: String Theory and the Geometry of the Universe's Hidden Dimensions. Basic Books, A Member of the Perseus Books Group (2010)
5. Kurosh, A.G.: The Theory of Groups. American Mathematical Society (2003)
6. Goldblatt, R.: The Categorial Analysis of Logic. Studies in Logic and the Foundations of Mathematics, vol. 98, revised ed., Elsevier Science Publishers, The Netherlands, Amsterdam XVI (1984)
7. Raikov, A.N.: Convergent Cognitype for Speeding-up the Strategic Conversation. In: IFAC Proceedings Volumes, vol. 41(2), pp. 8103–8108. Seoul, South Korea (2008). https://doi.org/10.3182/20080706-5-KR-1001.01368
8. Zhang, D.: Fundamentals of Image Data Mining Analysis, Features, Classification and Retrieval. Federation University Australia, Churchill, Australia. Springer, Cham (2019). https://doi.org/10.1007/978-3-030-17989-2

Chapter 15
Thought Boosts

Following [1], let us assign to each ordered pair of points (P, Q) of the Minkowski space $\mathcal{M}$ a vector interpreting some thought. A thought carries a definite meaning, i.e., it can be reflected in other thoughts, as well as on some events, phenomena, and relations between them. Moreover, events may differ in the level of certainty, be from different signal systems (e.g., mental and intuitive systems), exist in different places in space (brain, galaxy, etc.), and be directed. This vector representation of thought events can be fruitful.

Thoughts are in motion: they change, add up, intensify, and weaken. The vectors corresponding to them can be shifted, added, and multiplied. A human usually does this in the course of thinking, considering some thoughts together or making some of them more or less important (significant). With an algebraic interpretation of thought processes, axiomatic requirements for the transformation of vectors are possible:

(a) the associativity of addition, which determines the change in the distribution of attention to events in the sequence of reasoning;
(b) the commutativity of addition, i.e., the ability to change the sequence of understanding events;
(c) the zero vector, which characterizes a thought but adds nothing to it;
(d) replacing the product of the sum of the significance coefficients and a thought by the sum of the products of each coefficient and the thought;
(e) a unit event, which does not change any thought when multiplied by it.

At the same time, note that if one admits the nonlocality of events, then the requirement of commutativity may not be satisfied. The resulting vector space, most likely, should have a minimum dimension of 4 for covering at least the space $\mathcal{M}$. Vectors in this space can be represented as a linear combination of linearly independent vectors $(e_0, e_1, e_2, e_3, \ldots)$ that make up its basis. In this case, a normalized basis with the so-called *contravariant* coordinates can be considered:

$$e_0 = 1, \quad e_1 = e_2 = e_3 = -1. \tag{15.1}$$

© The Author(s), under exclusive license to Springer Nature Singapore Pte Ltd. 2021
A. Raikov, *Cognitive Semantics of Artificial Intelligence: A New Perspective*,
SpringerBriefs in Computational Intelligence,
https://doi.org/10.1007/978-981-33-6750-0_15

In this space, with any two vectors a real number can be associated, i.e., the standard *scalar product* can be defined. It satisfies the axioms of linearity and symmetry, and it can take positive, zero, and negative values. This scalar product is needed to rank thoughts.

It is often important to introduce the concept of an interval between points (thought events), defined as the number

$$I(P, Q) = (\overrightarrow{PQ}, \overrightarrow{PQ}). \tag{15.2}$$

The interval determines a quantitative characteristic (value) for the meaning of a thought since meanings are usually associated with a pair of events and with the paradoxes of their relationships. If this value is equal to 0, then the thought withdraws into itself and becomes closed; it has not been revealed so far. If a thought is attributed to something else, e.g., an external object or an internal intuitive event, then the interval takes a nonzero value. If the interval has a positive value, then the thought exists and changes in the traditional time; if the interval has a negative value, then it characterizes an event from the past that can be represented in a multidimensional spatial plane.

The basis (15.1) can be used for representing vectors. A given vector x is expanded with respect to this basis through its coordinates x^α:

$$x = x^\alpha e_\alpha. \tag{15.3}$$

In other words, the scalar product of vectors can be written through their coordinates in a standard way. Then the numbers

$$x_\alpha = (x, e_\alpha) \tag{15.4}$$

are called the coordinates of the vector x in the basis e_α.

Fixing a point O in the space $\mathcal{M}$, with keeping the same intervals between the points, corresponds to the so-called Lorentz transformation Λ. Such a technique "attaches" the transformation to a definite thought event, subject, or group of people. That is,

$$I(\Lambda P, \Lambda Q) = I(P, Q). \tag{15.5}$$

Replacing P and Q by the corresponding vectors, x and y, with the endpoints at the origin, let us write

$$(\Lambda x, \Lambda y) = (x, y). \tag{15.6}$$

The Lorentz transformation preserves the scalar product. Due to (15.6), the transformation of vectors is *linear*. From the standpoint of interpretations of thought processes, this property is admissible since the Lorentz transformation to nonlinear processes can be extended, e.g., by interpolating a nonlinear function and using the Fourier transform. Then the Lorentz transformation converts each vector of the space $\mathcal{M}$ into another vector.

Similarly, the Lorentz transformation can be applied to the quantum state (13.1), representing this state in the form of a set of vectors: $|u_1\rangle \to x$, $|u_2\rangle \to y$. Then

$$\Lambda(\alpha_1|u_1\rangle + \alpha_2|u_2\rangle) = \Lambda(\alpha_1 x + \alpha_2 y) = \alpha_1\Lambda x + \alpha_2\Lambda y. \tag{15.7}$$

The scalar product of vectors can be expressed through contravariant coordinates by introducing a metric tensor:

$$g_{\alpha\beta} = (e_\alpha, e_\beta), \tag{15.8}$$

$$g_{00} = 1, g_{11} = g_{22} = g_{33} = -1, \quad g_{\alpha\beta} = 0(\alpha \neq \beta).$$

For Λ to be the Lorentz transformation (see (15.6)), additional conditions have to be imposed on it as follows:

$$(\Lambda x, \Lambda y) = g_{\alpha\beta}(\Lambda x)^\alpha(\Lambda y)^\beta = g_{\alpha\beta}\Lambda_\gamma^\alpha\Lambda_\delta^\beta x^\gamma y^\delta, \tag{15.9}$$

must be equal to

$$(x, y) = g_{\gamma\delta}x^\gamma y^\delta,$$

where γ and δ are some indices.

Thus,

$$g_{\gamma\delta} = g_{\alpha\beta}\Lambda_\gamma^\alpha\Lambda_\delta^\beta.$$

Due to (15.7), the matrix (Λ), where the first index in $g_{\alpha\beta}$ corresponds to row number and the second to column number, defines the Lorentz transformation if and only if

$$\Lambda_\alpha^0\Lambda_\beta^0 - \Lambda_\alpha^1\Lambda_\beta^1 - \Lambda_\alpha^2\Lambda_\beta^2 - \Lambda_\alpha^3\Lambda_\beta^3 =$$
$$= \begin{cases} 0(\alpha \neq \beta) \\ 1(\alpha = \beta = 0) \\ -1(\alpha = \beta = 1, 2, 3) \end{cases} \tag{15.10}$$

The solution of this equation for $\Lambda_\beta^\alpha(\alpha, \beta = 0, 1)$ can be written through hyperbolic functions, e.g.,

$$\Lambda_0^0 = -\text{ch}\,\vartheta, \Lambda_1^0 = -\text{sh}\,\vartheta, \Lambda_0^1 = -\text{sh}\,\vartheta, \Lambda_1^1 = -\text{ch}\,\vartheta,$$

where ϑ is a real parameter. For the transformations with $\Lambda_0^0 > 0$ and det $\Lambda > 0$, let

$$\text{ch}\,\vartheta = \frac{1}{\sqrt{1 - v^2/c^2}}, \quad \text{sh}\,\vartheta = \frac{-v/c}{\sqrt{1 - v^2/c^2}}, \quad x^0 = ct, \quad x^1 = x,$$

which finally gives the well-known Lorentz transformations

$$x' = \frac{x - vt}{\sqrt{1 - v^2/c^2}}, \quad t' = \frac{t - vx/c^2}{\sqrt{1 - v^2/c^2}}. \tag{15.11}$$

The Lorentz transformations make up the Lorentz group. In the theory of relativity, they interpret the transformations of thought events in the spatiotemporal coordinates of the space $\mathcal{M}$ during the transition from one inertial reference system to another. Thus, using group theory for the interpretation of thought processes, through their vector representation in the form of a set of events, one simultaneously encompasses both the quantum-wave aspects of the microworld and the relativistic aspects of the macroworld.

On the one hand, the thought phenomenon is bound to some subject (including groups); on the other hand, it captures distant objects in space. In the Lorentz transformation Λ, the former feature is fixed by the origin of *the reference system O* and the basis (e_α), and the latter feature by assigning to each point (event) P of the space $\mathcal{M}$ some spatiotemporal coordinates (x^0, x^1, x^2, x^3) by the formula

$$\overrightarrow{OP} = x^\alpha e_\alpha. \tag{15.12}$$

In this reference system, all thought events are treated from the viewpoint of some observer, whose moments of life are determined by the coordinates x^1, x^2, and x^3. In this space, the vectors being multiples of e_0 form the time axis.

The transformations preserving the time axis, i.e., such that $\Lambda e_0 = e_0$, represent the rotations of the Euclidean space, making up a subgroup of the Lorentz group. In this subgroup, each rotation can be continuously transformed into another, which resembles thinking, which translates one state of thought into another (during a discussion, conversation). In this case, the instant of thought insight may occur, which suddenly comes after long reflections [2]; also, see Chap. 7. However, the resulting insight can be subsequently substantiated, and some basis can be assigned to it, in a continuous or logical way (deduction, induction, or abduction), using existing knowledge. Such a subgroup can be called *connected*.

The other part of the Lorentz group, which does not belong to a connected subgroup, is not related to the latter via any continuous changes. However, this separate part can itself be connected, i.e., the group allows some kind of clustering into connected subgroups. Note that this property of the Lorentz group resembles compactness in a topological space, which can be useful for ensuring a continuous convergence of inverse problem-solving in the course of thinking [3]; also, see Chap. 18. In this case, one of the connectivity components can preserve intervals and rotate Λ, whereas the others can not only preserve intervals (likewise) but also transform the values of the coordinates of events. At the same time, thinking is usually associated with some topic or direction of research carried out in time. This direction is also an event that needs to be fixed in the event space somehow. The concept of a *boost* [4] may be suitable here.

Let $\tilde{e}$ be an arbitrary nonzero vector from the basis (e_1, e_2, e_3) of the space $\mathscr{M}$. A boost in this space is an orthochronous transformation that does not change the vectors orthogonal to $(e_0, \tilde{e})$. (Recall that an orthochronous transformation preserves the direction of the time axis, in the case under consideration, e_0.) Then, Λ can be unambiguously decomposed into the product of a boost B and a *rotation R* that translates the basis (e_1, e_2, e_3) into the basis of the same orientation:

$$\Lambda = BR. \tag{15.13}$$

Thus, by specifying a *boost* and a rotation matrix, one can transform thought events in different ways, taking into account different classes of transformations (including the ones by the Lorentz formulas (15.11)) and the connected components of the Lorentz group. For accelerating the creative thought process, the transformation itself must be purposeful and stable, whereas the events can behave in a chaotic manner. The chaotic interpretation of the purposeful thought process is discussed in detail in Chap. 16.

References

1. Rumer, Yu.B., Fet, A.I.: Teoriya Grupp i Kvantovye Polya (Group Theory and Quantum Fields). Russia, Moscow, LIBROKOM (2007)
2. Perkins, D.: The Eureka Effect. The Art and Logic of Breakthrough Thinking. Norton, New York, London (2001)
3. Ivanov, V.K.: Incorrect Problems in Topological Spaces. Siberian Math. J. 10, 785–791 (1969). Novosibirsk, Russia
4. Ramadevi, P.: Dubey, V: Group Theory for Physicists, with Applications. Cambridge University Press, UK (2019)

Chapter 16
Purposeful Chaos

According to the aforesaid, a thought can be represented as a set of spatiotemporal events: these are a wave and particles that evolve in a certain direction and in a chaotic manner. While a thought has not yet reached the "logical surface" (i.e., not expressed in words), the thought process is chaotic and limited by a conditional boundary dictated by the subject of thought. The purpose of thinking is to bring order within the chaos. At the same time, if this order is introduced rigidly and taking the existing knowledge into account, then the expected thought insight will obviously not occur since the thought will be squeezed in the blinders of the established logical stereotypes. As a result, only deduction, induction, or abduction will work. If one puts no pressure on the thought, giving complete freedom, then it will most likely be scattered, become divergent, and, possibly, will be lost.

The idea put forward in this book is that a thought will, sooner or later, concentrate on the solution of some problem and the achievement of some goal. In other words, its behavior will be convergent [1]; also, see Chap. 8. Then the environment in which a thought develops can become relatively homogeneous, dense, and, possibly, incompressible. This means that divergence can be conditionally equated to 0. Sooner or later, a thought process ends; at this time, there will be a kind of damping of its chaotic oscillations. Note that the fundamental laws of thermodynamics must anyway hold.

In a somewhat conditional way, let a thought in the spatiotemporal interpretation be denoted by a function $U(z, t)$, where z is a place in space and t is time. When seeking a solution of some problem, a thought can obey the bizarre "Eureka" effect (also, see Chap. 7 and [2]), and its behavior can be illustrated by Fig. 16.1.

Of course, a thought cannot be directly described by a formula or words because otherwise, it would immediately cease to be a thought, turning into a communication scheme. Verbalization will neglect the potential of emotions and the transcendental state of mind. However, various laws of physics and mathematics can help describe and represent a chaotic thought indirectly. For example, the behavior of a thought can be wavy, abrupt (jump-like), quantized, chaotic, etc. The wave nature of a thought

A. Raikov, *Cognitive Semantics of Artificial Intelligence: A New Perspective*,
SpringerBriefs in Computational Intelligence,
https://doi.org/10.1007/978-981-33-6750-0_16

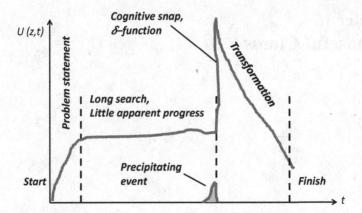

Fig. 16.1 "Eureka" effect

has been illustrated using the d'Alembert equation (see Chap. 12) and the quantum one (see Chaps. 10, 13, and 21) through the effect of nonlocality. Now reflect the chaotic peculiarity of a thought by referring to the representation of a corresponding dynamic system.

The Lagrangian can be used to describe the state of a dynamic system, as has been mentioned above in relation to quantum fields; see Chap. 14. The Lagrangian describes the states of a quantum field, has an infinite number of degrees of freedom, and is invariant with respect to the local transformations of groups. The Lagrangian can be chosen in various ways.

The interaction of thought events, represented in a dynamic system, e.g., in the form of some multidimensional space in group theory, depends on distances or intervals. In addition, one should take into account the paradoxes of thinking, such as the effect of quantum entanglement (nonlocality). In each pair, two events are linked to one another, forming connected Lorentz subgroups: one rotation is translated into another by a continuous change. With this interpretation, the state of a thinking system is specified by a set of positions and vectors of events, e.g., in the system of generalized coordinates $\left(x^0, x^1, x^2, x^3\right)$ and velocities $\left(\dot{x}^0, \dot{x}^1, \dot{x}^2, \dot{x}^3\right)$ of the space $\mathcal{M}$,

or by vectors $\mathbf{q}_i (i = 0, \ldots n)$ and $\dot{\mathbf{q}}_i (i = 0, \ldots n)$ in an infinite-dimensional Hilbert space. Changes take place in these spaces: states are turning one into another over time. These changes can be defined:

(a) in a formalized way, using Hamilton's principle of least action;
(b) in an unformalized way, with assessing the meaningfulness of the action by an external observer.

In either case, one needs a function determining the state of the system. In the case of mathematical and physical interpretation of changes, this can be the Lagrangian L. (The symbol should not be confused with the Laplace transform.) For the time being, the form of the Lagrangian L will not be specified, if only because a subjective factor can be included in the process. Note that it depends on the variables q_i, $\dot{q}_i$, and t. After all, the truth requires a definite localization of events.

In case (a), according to Hamilton's principle, it is necessary to minimize *the action integral*

$$W = \int_{t_1}^{t_2} L\left(\mathbf{q_i}, \dot{\mathbf{q_i}}, t\right) dt. \tag{16.1}$$

In case (b), an external observer intervenes in the decision process: using his or her knowledge and feelings, the external observer assesses the result of a (final or intermediate) decision and gives his or her opinion, including the completion of the decision. Thus, the inverse problem can be solved when the target (terminal) value of the path is inexactly known. Moreover, it may be unknown whether there exists any solution at all and whether it is unique. In other words, case b) considers an ill-posed problem in some conceptual space. The solution will be discussed below; see Chap. 18.

Assume that the states of the system at time instants t_1 and t_2 are known, and the path must be found. This is the classical formulation of an optimization problem in mechanical systems: find an optimum of the path function between the points $\mathbf{q}(t_1)$ and $\mathbf{q}(t_2)$, the initial and terminal points of the path, respectively. In the case under consideration, both the path and the corresponding function are unknown. The terminal point of the path is also inexactly known; there is only an intention, a desire, and a direction of action. For example, scientific research can be carried out with some problem formulation that is unsolvable at the current time. For example, a problem is to find physical evidence of cosmic strings or to elaborate a theory of quantum gravity. Therefore, before calculating the optimum of a function, it is necessary to guess this function and the terminal point of the path hypothetically.

The optimum of a function can be obtained by introducing small deviations of its values along the path, which vanish at the ends of the path: $\delta\mathbf{q}(t_1) = \delta\mathbf{q}(t_2) = 0$. For the sake of clarity, assume that the degree of freedom is 1:

$$\delta W = \int_{t_1}^{t_2} L_i\left(q_i + \delta q_i, \dot{q}_i + \delta \dot{q}_i, t\right) dt - \int_{t_1}^{t_2} L\left(\dot{q}_i, \dot{q}_i, t\right) dt. \tag{16.2}$$

Let us expand the integrand into a Taylor series and, setting the left-hand side of equality (16.2) equal to 0 (the optimum of the function), write the following equation for the Lagrangian L:

$$\frac{dL}{dq} - \frac{d}{dt}(dL/d\dot{q}) + \delta\varphi = 0, \tag{16.3}$$

where $\delta\varphi$ is the remainder of the series containing the terms of higher orders, starting with the second order. When determining the form of the function $f(x)$ representing the Lagrangian L, this remainder is usually omitted. At the same time, such a simplification may turn out to be unacceptable when using AI to analyze infinitesimal values of data or signals, e.g., coming from the cosmos or a quantum particle accelerator.

Also, the function can be of a complex variable z and considered in the neighborhood of some point a. In contrast to the real variable case, under the Cauchy–Riemann conditions satisfied in some neighborhood of the point a, the convergence of the series to the function $f(x)$ can be examined. If $\delta\varphi = 0$ (the higher-order terms are neglected), then this useful case becomes degenerate.

Note that when $\delta\varphi$ vanishes, a large class of problems in which the field (signal) sources can be written in the form of the Fourier integral over their harmonic components has no rigorous solution. According to formulas (16.2) and (16.3), the analog (non-discrete) signal becomes a complex number and can be represented by adding to the original function an imaginary part formed using the Hilbert transform. In this case, the direct and inverse problems in a wave medium are solved using the d'Alembert equations. The solution of this equation is represented as a sum of traveling waves; or, when using the Fourier expansion, as an infinite linear combination of harmonic functions. Therefore, in applications, neglecting $\delta\varphi$ may turn out to be unacceptable. Perhaps, due to the disregard of this nuance when using deterministic AI tools based on digital data and their interpolations, AI is viewed as "insufficiently powerful" for studying the phenomena mentioned above, such as cosmic strings or particle collisions in the LHC.

Thus, given n degrees of freedom, the standard Lagrange equations are obtained by varying each variable independently in the form of Eq. (16.3) and letting $\delta\varphi = 0$. Then, for the mechanical case, the Lagrangian can be represented as the difference between the kinetic (K) and potential (U) energies of the particles forming the system [1]. When elements of a phenomenological nature are included in the process, such "mechanical" reductions have very conditional applicability.

This interpretation can be improved by involving the Hamilton equations, which better suit the study of chaotic and statistical phenomena, as well as quantum mechanics. Although, in any case, such an approach to the representation of thought processes remains conditional. Addressing such formalisms can be useful for interpreting, e.g., the quantum nonlocality of thinking, or even for investigating, in this nonlocal context, the relationship of thought processes with cosmological phenomena such as cosmic strings. In particular, the such phenomena are investigated using the methods of smooth manifolds [3, 4], which makes it possible to model individual thought events of nonlocal nature using differential equations and, consequently, stability theory based on Lyapunov functions.

Taking advantage of such associations, represent a dynamic Hamiltonian system as a set of components differing by homogeneous properties [5]. Consider Eq. (16.3) and let $\delta\varphi = 0$. Then, for the classical Hamiltonian system, there exists a Lagrangian $L = K - U$, where K and U denote the kinetic and potential energies, respectively. As a result, the dynamics of the situation can be described using the Lagrange equation, and its stability can be analyzed using a Lyapunov function V equal to the total energy of the system, $E = K + U$.

Consider the thought process of solving some problem and include in it all possible elements:

- of deterministic nature (words, images, formulas, structures, and schemes);
- of phenomenological nature (feelings, emotions, free will, and nonlocal impacts).

In other words, take into account all external and internal impacts. Then, all external impacts can be incorporated into this thought system by making it isolated from the external environment (closed). Next, recall the well-known fact of thermodynamics: in a closed system, the entropy S grows. At the same time, according to the law of conservation of energy, $E = V = const$.

Quite obviously, in a chaotic system, the production of entropy and the Lyapunov function V satisfy the relation

$$\sigma = dS/dt = -(1/T)dV/dt, \tag{16.4}$$

where S denotes the system's entropy, $T > 0$ is a normalizing factor, and σ is the rate of entropy generation. According to this relation, stability cannot be achieved in a non-equilibrium closed system since $\sigma > 0$ and $dV/dt = 0$ due to $V = const$.

Thinking intended for problem-solving produces hypotheses, and new thoughts in a closed system increase its entropy and create a new motion. At the same time, a change in the elements of a deterministic nature generates and fixes knowledge as well as provides communication, thereby creating an order, e.g., in the form of knowledge bases, action patterns, and algorithms. For such a closed system, the second law of thermodynamics must be satisfied, i.e., $\sigma > 0$. Hence, for the relation (16.4), the Lyapunov stability conditions hold only if $dV/dt < 0$, which is impossible for a closed system. In a closed system, uncertainty is constantly growing, the conditions of stability are violated, and the system actually degrades.

Thus, to ensure the stability of any system with a source of chaos, one has to remove isolation and "open" the system, allowing information exchange between the internal and external environments. Such an "opening" can be implemented by extracting from the system a chaotic source of information, or some force, and placing the latter into the external environment as an independent system. It can be an individual or a group with emotions, feelings, and desires. As was shown in [5], the stable behavior (development) of the former system can be defined by the relation

$$dV/dt = P*P' + (S_{int} - S_{exch}) * \left(S'_{int} - S'_{exch}\right), \tag{16.5}$$

where P and P' are the level and rate of establishing order in the system (the parameters of a deterministic nature); S_{int} and S'_{int} are the level and rate of growing internal disorder in a thought process (the parameters of a phenomenological nature); finally, S_{exch} and S'_{exch} are the level and rate of exchanging information (chaotic external information coming into the system).

This formula can be used for achieving sustainable development in a thought process via an appropriate balance for the chaotic and logical elements of the system. At the same time, additional conditions are needed for making this process purposeful. They can be found using inverse problem-solving methods; see Chap. 18.

In this chapter, the case of chaotic thinking without dissipation has been considered. The phenomenological component of thinking has been interpreted using Hamilton's equations and elements of Lagrangian mechanics. Within the Lagrangian approach, the system under study has been characterized by the function of generalized coordinates, velocities, and time. The Hamiltonian approach has involved the concept of generalized momenta conjugate to the generalized coordinates and defined through the Lagrangian. In this case, in a scalar field, the velocity has been characterized not by vector directivity but by its scalar potential only. In the next chapter, the aspect of dissipation will be introduced into the models of thought processes.

References

1. Raikov, A.N.: Convergent Cognitype for Speeding-up the Strategic Conversation. In: IFAC Proceedings Volumes, vol. 41(2), pp. 8103–8108. Seoul, South Korea. https://doi.org/10.3182/20080706-5-KR-1001.01368 (2008)
2. Perkins, D.: The Eureka Effect. The Art and Logic of Breakthrough Thinking. Norton, New York, London (2001)
3. Sazhina, O.S., et al.: Optical Analysis of a CMB Cosmic String Candidate. Adv. Access Publ. MNRAS **485**, 1876–1885 (2019). https://doi.org/10.1093/mnras/stz527
4. Yau, S.-T., Nadis, S.: The Shape of Inner Space: String Theory and the Geometry of the Universe's Hidden Dimensions. Basic Books, A Member of the Perseus Books Group (2010)
5. Ulyanov, S.V., Raikov, A.N.: Chaotic Factor in Intelligent Information Decision Support Systems. Aliev, R., (Ed.). In: Proceedings of the 3rd International Conference on Application of Fuzzy Systems and Soft Computing (ICAFS'98). Wiesbaden, Germany, pp. 240–245 (1998)

Chapter 17
Dissipative Thought

When considering the dynamic processes of chaotic thinking without dissipation (see the previous chapter), the problem is only to ensure the stability of the process. Here the conditions of balanced control preserving the zero or any negative value of its divergence have to be found.

At the same time, a natural assumption is that the solution of any problem will be completed sooner or later. If the thought process is damping (the problem is being solved), it becomes dissipative. In this case, the process may have oscillations, chaotic behavior, and turbulence effects. The process is under pressure from external circumstances: it is influenced by signals coming from various sources of information. Even in classical physics, problems under such conditions as dissipation, diffraction, and turbulence, are difficult to solve. Such problems are often considered and solved under restrictions and assumptions.

Note that the environment (medium) in which thinking is carried out can be treated incompressible. This medium is characterized by both wave and chaotic behavior. The fields and the wave nature of signal propagation have been described above using such fundamental principles as the d'Alembert and Helmholtz equations and the Green function; see Chap. 12. For describing the chaotic characteristics of thinking, the idea is to try employing the Navier–Stokes equation [1] for vector fields, with the subsequent association and integration using the d'Alembert equation. For an incompressible medium, the Navier–Stokes equation has the form

$$\frac{\partial u}{\partial t} + (u\nabla)u - \nu\nabla^2 u = \rho\nabla p + \mathbf{f}, \tag{17.1a}$$

$$\mathrm{div}\, u = 0, \tag{17.1b}$$

$$u = 0 \text{ on } \mathscr{B}, \tag{17.1c}$$

© The Author(s), under exclusive license to Springer Nature Singapore Pte Ltd. 2021
A. Raikov, *Cognitive Semantics of Artificial Intelligence: A New Perspective*,
SpringerBriefs in Computational Intelligence,
https://doi.org/10.1007/978-981-33-6750-0_17

where ∇^2 denotes the Laplace operator (also, see the d'Alembert Eq. (12.1)); $\mathscr{B}$ is a conditional boundary of the domain covering a thought process; u is a vector field of the velocities of thought flows (in contrast to the d'Alembert equation, here a vector variable); p gives pressure in the domain with the boundary $\mathscr{B}$; ρ is the density of the medium; f are external forces; finally, ν denotes the kinematic viscosity of the medium (m^2/s). The latter characteristic also depends on the dynamic viscosity of the medium (kg/m·s).

Consider a thought process leading to insight; see Fig. 16.1. The process is initiated and gradually growing, without any result for a long time. At some time instant, a solution suddenly comes. The instant of insight can be represented as the δ-function, which is equal to infinity, but the integral of which is equal to 1.

Note that for transforming a scalar field into a vector field, one needs to find the gradient. This is the simplest method of such a transformation. For this purpose, the Hamilton operator is often used. It represents a differential vector-valued operator whose components are partial derivatives with respect to the corresponding coordinates:

$$\text{Grad}\, u = \nabla u. \tag{17.2}$$

As is known, a vector field where divergence (17.1b) vanishes at all points is called a solenoidal vector field. It can be represented as the curl of some other vector field. The transformation of a vector field into a scalar field will require replacing the curl with some scalar value or considering it equal to 0. A vector field where the curl vanishes at any point is called potential (non-vortex, irrotational). Such a field can be represented as the gradient of some scalar field (potential).

At the same time, the well-known Helmholtz theorem states the following: if at all points of a certain domain (e.g., with a boundary $\mathscr{B}$) a vector field has a divergence and a curl, then it can be represented as a sum of potential (scalar) and solenoidal vector fields. A vector field where both the divergence and the curl vanish at all points is called harmonic, and its potential is a harmonic function. Recall that in Eq. (17.1a), u is the vector field of velocities, and the divergence vanishes. Hence, it is possible to replace the velocity vectors temporarily with their scalar potentials, like in the scalar d'Alembert Eq. (12.1). Then, the Navier–Stokes equation for the velocity field u can be written in the scalar form:

$$\frac{\partial u}{\partial t} + (u\nabla)u - \frac{1}{R}\nabla^2 u = -\nabla p + \mathbf{f}, \tag{17.3}$$

where t denotes time; ∇ is the nabla operator (the gradient whose components are the partial derivatives with respect to different coordinates); R is the Reynolds number; p is the "dimensionless" pressure; finally, $\mathbf{f}$ is the vector field of a body force (the force acting on each elementary volume of a substance and proportional to the mass of this substance). The number R is given by

$$R = \frac{\rho V c_0}{\omega b},\qquad (17.4)$$

where ρ denotes the density of the medium (kg/m^3); V is the amplitude of the oscillating velocity (m/s); ω means the angular frequency (rad/s); c_0 is the velocity of signal propagation (m/s); b is the parameter of dissipation (viscosity—the property of a fluid body to resist the displacement of one its part relative to another; due to this property, the work expended on such a displacement is dissipated in the form of heat). This number may vary within large ranges, from 20 to 10000.

For each type of fluid (medium, thought), there exists a critical value of the Reynolds number, which determines the transition from laminar flow to turbulent flow.

Note that the relation $\frac{\partial u}{\partial t} = 0$ holds on the time interval of the process without any significant progress in the solution (see Fig. 16.1); without loss of generality, let $\frac{\partial u}{\partial t} = const \neq 0$ on the time interval of the process with initiated problem-solving.

Now suppose that the velocity potential at the points in the space u (see Eqs. (12.1) and (17.3)) has the same values. Substituting the value $\nabla^2 u$ from (12.1) into Eq. (17.3), write the equation

$$\frac{\partial u}{\partial t} + (u\nabla)u - \frac{1}{R}\left(\frac{1}{c(x)^2}\frac{\partial^2}{\partial t^2}u + F(x,t)\right) = -\nabla p + \mathbf{f}.\qquad (17.5)$$

In this case, the external force influencing the process is represented in this equation by the signal source $F(x,t)$ and by $\mathbf{f}$ (if any). The signal source can have an interpretation of some force initiating the process of problem-solving. The gradient operator ∇ of the medium provides information about the vector directivity of the process; therefore, the vector character of $\mathbf{f}$ can be conditionally omitted. Then both quantities $F(x,t)$ and $\mathbf{f}$ can, up to a certain coefficient, be combined into one scalar quantity, further denoted by f'.

Equation (17.5) is valid for any point in the space outlined by the boundary $\mathscr{B}$ (which can be quite large), both with respect to the d'Alembert equation and the Navier–Stokes equation. Then $R = const$, and the propagation velocity c (of signal, information, sound, or thought) begins to coincide with the velocity u. After small transformations, this equation takes the form

$$R\frac{\partial u}{\partial t} + R(u\nabla)u - \frac{1}{u^2}\frac{\partial^2}{\partial t^2}u = -R\nabla p + f'.\qquad (17.6)$$

As is well-known, the Reynolds number measures the ratio of the inertial forces acting in the flow to the forces of viscosity. The density ρ in the numerator of formula (17.4) characterizes the inertia of particles with acceleration, and the value of viscosity in the denominator characterizes the tendency of the medium to impede such acceleration. Also, the Reynolds number can be considered as the ratio of the kinetic energy of the signal, etc., to the energy loss on a typical length (due to internal friction, aberration, diffraction, or interference in the medium).

If the Reynolds number of a flow is many times greater than the critical one (under which the flow changes its state from laminar to turbulent), then the medium can be considered ideal. In this case, the viscosity value can be neglected since the thickness of the boundary layer is small compared to the characteristic size of the process, i.e., the viscous forces are significant only in a thin layer (developed turbulence is observed in the flow). Taking into account that $R = \text{const}$, Eq. (17.6) reduces to a standard form by letting $R = 1$.

Thus, a thought flow can be described by the equation

$$\frac{\partial^2}{\partial t^2}u - u^2\frac{\partial u}{\partial t} - \left(u^3\nabla\right)u\nabla p + \mathbf{f}' = 0. \tag{17.7}$$

This equation is not strict for thought processes but determines only the behavioral tendencies of a dissipative, chaotic (turbulent), and wave-like thought.

Obviously, the solution of this equation is far from simple, if even exists. For example, it seems interesting to solve (17.7) at the point of insight, when u behaves like the δ-function. At this time instant, the integral of the values of the function u is equal to 1. Another approach is to find a solution on a "plateau" where the derivative of u vanishes. For such points, the equation can take a particular (and possibly simpler) form. At the same time, it makes little sense to solve this equation: thinking cannot be replaced by a formula, as is repeatedly mentioned throughout the book. Well enough, this equation conventionally shows the behavioral dynamics of a thought. For example, the constant velocity value affects the thinking process with a negative sign. The value of the gradient of the velocity and pressure of the medium, which can be interpreted, e.g., as a directed (vector) pressure imposed by the inertia of thinking, as well as by the stereotypes of thinking or conformism, respectively, has a negative effect. The presence of some force, which is interpreted by the source of information, has a positive effect on a thought process. In particular, such a force can be information about the goal of thinking; such information makes thoughts purposeful.

Reference

1. Campos, D., (Ed.): Handbook on Navier–Stokes Equations: Theory and Applied Analysis. Physics Research and Technology. Nova Science Publishers, New York, USA, p. 600 (2017)

Chapter 18
Purposefulness of Thought

Thinking does not fit into formal and mathematical schemes. This activity operates with the uncaused, non-quantitative, and nonmetric spaces rather than with the metric ones. Concepts in such spaces can be conventionally represented by "points," neighborhoods, and sets in the theory of groups, categories, and topologies [1, 2]. The problem itself can have an inverse character. For example, strategic planning is inverse: ambitious goals set by the top managers of a company lie beyond the extrapolation of the current dynamics of events. To achieve such goals, it is necessary to change the management paradigm, to find and assess unusual ways to succeed.

Consider a purposeful thought process of an individual or a group with inexact goals. Such a process can be represented as solving an inverse problem on nonmetric spaces X and Y:

$$x = Q^{-1} y_\delta, \tag{18.1}$$

where $x \in X$ denotes resources (means) for achieving the goal; $y_\delta \in Y$ is an inexact goal in the goals space Y; finally, Q^{-1} is an inverse operator characterizing the process of resource utilization for achieving the goal. Inverse problems answer the question: "What needs to be done to…?" Such problems are ill-conditioned and characterized by incorrectness (ill-posedness): they may have no solution; there may exist a set of solutions; minor variations in the initial conditions, the supply of new information, etc., may cause a significant change in the resulting solution.

Inverse problems can be solved, e.g., using topological spaces. In topology, one can operate with metrically immeasurable concepts, when the distances between the concepts (points) cannot be measured in a traditional way, and an unstable solution of the problem can be avoided only via human participation: individuals and groups introduce non-formalized qualitative information into the process of problem-solving. As soon as humans escape the decision process, the process collapses, i.e.,

© The Author(s), under exclusive license to Springer Nature Singapore Pte Ltd. 2021   93
A. Raikov, *Cognitive Semantics of Artificial Intelligence: A New Perspective*,
SpringerBriefs in Computational Intelligence,
https://doi.org/10.1007/978-981-33-6750-0_18

the process stops. The problem (18.1) can be formulated more rigorously as the goal-setting problem; for details, see Chap. 8. The classical approach is not suitable for solving such problems.

Suppose that there exists a unique solution of the problem. This assumption is quite natural for practice since a certain solution will eventually be found. In other words, the set of preimages $Q^{-1}V_\delta$ of all neighborhoods of the point y_0 (exact goal) has a single point x_0 (exact solution) at the intersection. However, this filter may have no convergence to the point x_0; in this case, the problem becomes incorrect, and special rules (conditions) are needed for involving a human in the problem-solving process.

Traditional search engines use the concept of relevance for measuring the quality of their operation. A formalized search query is created (e.g., words, diagrams, or pictures); in accordance with formal rules, the computer finds the corresponding entries in the data array. These formal rules can be logically complex and even based on neural networks. The entries are also described in a formal language. The user may be unsatisfied with the response of the system; he or she needs something else, though does not always understand what exactly. Anyway, the user assesses the response, lets the machine know the assessment, and the machine performs a new search. The developers of search engines always face the problem of speeding up such a procedure in which the search is not reduced to a formalized algorithm, and it is necessary to involve an unformalized subjective factor in the process.

Logical approaches cannot always be effective for creating a path to the goal $y_0 \in Y$ in inverse problem-solving, which is incorrect. Their formalizable nature predetermines the shortcoming of traditional approaches. When applied to describe human emotions [3, 4], logical modeling destroys its nature. The problem requires using a non-logical approach of infinite continual power. The occurrence of thought events can be unformalized and chaotic, and the goals can be ill-defined. This is an inverse problem, and its solution needs a special approach for ensuring the convergence of the process in a conceptual space.

Traditional logic-like AI has significant limitations for embracing the non-logical aspects of decision-making [5]. Category theory, topology theory, and inverse problem-solving will make AI stronger: these supplements to AI can be used to represent the unformalized processes of decision-making.

Inverse problem-solving is not stable, and the resulting solution may have significant changes under small variations in the source data. The convergent approach can be applied to solve such problems [6, 7]. Special mathematical conditions for making the process convergent are as follows: if X is a topological space and Y is a compact Hausdorff space, then the graph of A is closed if and only if A is continuous [8]. Figure 18.1 illustrates necessary conditions under which the solution of Eq. (18.1) is convergent.

The convergent approach provides a list of rules to structure information, as follows:

- A tree of goals should be created, and the goals should be ranked by importance (*purposefulness*).

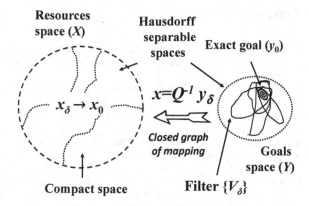

Fig. 18.1 Necessary conditions for convergent decision-making

- The semantic interpretations of each pair of points (events) should be *separable* (*Hausdorff space*).
- The resources space X should be represented by a finite and a foreseeable set of elements (*compact space*).
- Decision-makers should introduce non-formalized qualitative information in the problem-solving process (*closed graph of mapping*).

These rules ensure the purposefulness and stable convergence of the decision-making process to the ill-defined goals. The participants repeatedly introduce qualitative information into the process during computer simulations, thereby ensuring that the process evolves towards the goal.

Qualitative information introduced into the decision-making process lies beyond the framework of mathematical logic and formalisms and can be dictated by the individual capabilities and circumstances, experience and temperament, dignity and preferences, faith and patience, importance and caution, simplicity and immensity, isolation and impenetrability, timidity and, possibly, confusion of different participants of the decision-making process.

Sometimes there is not enough initial data to structure information in accordance with the listed rules. For example, the group may not yet know their goals. However, group members usually have the intention and act accordingly. In this case, the behavior of an individual or a group of people shows a tendency to analyze and systematize information: information is naturally classified and decomposed on some basis into a finite number of parts. So, in the conditions of uncertain goals, the objects capable of self-organization are divided into groups, categories, and clusters (compact space interpretation).

Nevertheless, the enlargement of all components of a thought process in decision-making through the identification of such categories does not remove the need to consider very small impacts on it.

References

1. Goldblatt, R.: The Categorial Analysis of Logic. Studies in Logic and the Foundations of Mathematics, vol. 98, revised ed., p. XVI. Elsevier Science Publishers, Amsterdam, The Netherlands (1984)
2. Wong, C.K.: Covering Properties of Fuzzy Topological Spaces. J. Math. Anal. Appl. **43**, 697–704 (1973)
3. Perkins, D.: The Eureka Effect. The Art and Logic of Breakthrough Thinking. Norton, New York, London (2001)
4. Gigerenzer, G.: Gut Feelings. The Intelligence of the Unconscious. Viking, London (2007)
5. Raikov, A.N.: Holistic Discourse in the Network Cognitive Modeling. J. Math. Fundam. Sci. **3**, 519–530 (2013)
6. Raikov, A.: Post-non-Classical Artificial Intelligence and its Pioneer Practical Application. Part of Special Issue. In: 19th IFAC Conference on Technology, Culture, and International Stability (TECIS 2019), Sozopol, Bulgaria, IFAC-PapersOnLine, vol. 52(25), pp. 343–348 (2019). https://doi.org/10.1016/j.ifacol.2019.12.547
7. Raikov, A.N.: Convergent Cognitype for Speeding-up the Strategic Conversation. In: IFAC Proceedings Volumes, vol. 41(2), pp. 8103–8108. Seoul, South Korea (2008). https://doi.org/10.3182/20080706-5-KR-1001.01368
8. Ivanov, V.K.: Incorrect Problems in Topological Spaces. Sib. Math. J. **10**, 785–791 (1969)

Chapter 19
Small Impacts

The thought process can be subjected to seemingly very small impacts, e.g., an external hint causing insight; see Chap. 16. Perhaps these impacts include random deviations at the neural level or quantum fluctuations.

Suppose that the thought process can be influenced by a quantum particle. Consider a particle with a certain mass m, moving under the influence of some potential $V = V(\mathbf{q})$, where $\mathbf{q}$ is the coordinate of the particle in the three-dimensional space. If the time coordinate t is taken into account, then the four-dimensional, e.g., Minkowski, space naturally arises. If small impacts are of a quantum-mechanical nature, then the well-known interpretations of the Schrödinger equation can describe the behavior of the thought process (or rather, its individual components) under such impacts:

$$i\hbar\frac{d}{dt}\psi(\mathbf{q}, t) = \widehat{\mathcal{H}}\psi(\mathbf{q}, t), \tag{19.1}$$

where $\hbar$ denotes the Planck constant divided by 2π; $\psi(\mathbf{q}, t)$ is a wave function; finally, $\widehat{\mathcal{H}}$ is the Hamilton operator given by

$$\widehat{\mathcal{H}} = -\frac{\hbar^2}{2m}\nabla^2 + V(\mathbf{q}). \tag{19.2}$$

With the action integral S introduced, the substitution

$$\psi(\mathbf{q}, t) = \exp\left\{\frac{iS(\mathbf{q}, t)}{\hbar}\right\} \tag{19.3}$$

allows transforming Eq. (19.1) to

© The Author(s), under exclusive license to Springer Nature Singapore Pte Ltd. 2021
A. Raikov, *Cognitive Semantics of Artificial Intelligence: A New Perspective*,
SpringerBriefs in Computational Intelligence,
https://doi.org/10.1007/978-981-33-6750-0_19

$$-\frac{dS}{dt} = \frac{1}{2m}\nabla S\,\nabla S + V(\mathbf{q}) - \frac{i\hbar}{2m}\nabla^2 S. \qquad (19.4)$$

The last term in this equation is very small, being often neglected accordingly. At the same time, the Planck constant represents a fundamental constant, and its appearance in this equation is not by chance. It can be significant. For this, the action integral S must be commensurable with $\hbar$. According to the substitution (19.3), letting $\hbar \to 0$ will infinitely accelerate the oscillations of the wave function. This leads to a *singular* effect, resembling the instant of insight. Recall that it has been written as the δ-function; see Fig. 16.1.

For estimating and finding the functional S, the last term in Eq. (19.4) can be temporarily neglected. Then it will be the Hamilton–Jacobi equation depending on the time variable and some value S_0, an approximation of S:

$$S_0(\mathbf{q}, t) = \int_{t_0}^{t} L(\mathbf{q}, \dot{\mathbf{q}}, t')dt', \qquad (19.5)$$

where $L(\mathbf{q}, \dot{\mathbf{q}}, t')$ denotes the Lagrangian; t_0 and t are the initial and terminal time instants, respectively.

For studying very small impacts on the thought process of decision-making, let $\hbar$ be a small parameter. Consider the one-dimensional time-independent Schrödinger equation:

$$-\frac{\hbar^2}{2m}\frac{d^2\psi}{dx^2} + (E - V(x))\psi = 0, \qquad (19.6)$$

where E denotes the stationary-state energy; $V(x)$ is a smooth potential by one of the coordinates x.

The approximate wave function can be found by substituting the one-dimensional version of (19.3) into (19.6), where $S(x)$ is the series

$$S(x) = S_0 + \hbar S_1 + \hbar^2 S_2 + \dots \qquad (19.7)$$

Based on the result of this representation, two solution domains are identified: $E > V(x)$ (in which the momenta $p(x)$ are positive) and $E < V(x)$ (in which the momenta $p(x)$ have imaginary values). The former belongs to the classically solvable domains; the latter corresponds to the domain through which the well-known quantum tunneling can occur [1]. This is a phenomenon when quantum particles can pass through energy barriers whose height exceeds the energy of the particles.

In the neighborhood of the points with $E = V(x)$ and $p(x) = 0$, the probability density $|\psi(x)|^2$ becomes very large. This agrees well with the idea that at such a point, the particle has a small momentum and moves slowly; hence, here, it can be fixed as easily as possible. At such a point, one should search for a solution under the following conditions: the transition from the domain with $p(x) > 0$ to the domain with imaginary values of $p(x)$ is smooth, and no divergence occurs. Then uniform approximations of solutions can be constructed. Similar conditions are satisfied, e.g., in geometric optics when studying caustics: rays are focused in the most bizarre way, forming very bright spots.

In geometric optics, a phase loss of $\pi/2$ can be observed when passing through a caustic. Moreover, to obtain solutions at such points, the classical action variables are assumed to be integer and multiples of $\hbar$, i.e., the values of the action variables are quantized; also, see Chap. 21. As is well known, there exist several quantization methods: Bohr–Sommerfeld, Einstein, and Einstein–Brillouin–Keller (EBK), to name a few.

The EBK quantization is used in multidimensional problems for making the wave function well-defined. More specifically, this method is applied to calculate eigenvalues in quantum mechanical systems. The pioneering theoretical results on determining the amplitudes of oscillations in the approximate wave function were obtained quite a long time ago; for example, see [2]. The EBK quantization is interpreted through the use of the action integral S, the substitution (19.3), and the immersion of the situation in a Hamiltonian system with N degrees of freedom.

To find the wave function $\psi(\mathbf{q})$ in this case, the general quantization condition can be written as

$$I_k = \oint_{\mathcal{E}_k} \mathbf{p}\,\mathbf{dp} = 2\pi\hbar\left[n_k + \frac{\alpha_k}{4}\right], \qquad (19.8)$$

with the following notations: I_k is the action variable corresponding to one of the degrees of freedom k; $\mathcal{E}_k$ is the constant energy curve; $\mathbf{p}$ is the momentum; n_k is the corresponding quantum number; finally, α_k is the number of crossed caustics. Note that the expression (19.8) reflects the integration over a closed circle of the constant energy curve. Caustics form quickly, with the velocity of light. A human thought is born relatively slowly, and insight requires time to accumulate knowledge. But the time lag can be filled with rapidly spreading waves of light and their quantization.

References

1. Polkinghorne, J.: Quantum Theory. A Very Short Introduction. Oxford University Press, New York (2002)
2. Van Vleck, J.H.: The Correspondence Principle in the Statistical Interpretation of Quantum Mechanics. Proc. Math. Acad. Sci. USA **14**, 178 (1928)

Chapter 20
Light Thought

Light—particles and waves—travels rapidly, but the distances can be very large, requiring billions of light-years. But the photon, as a quantum carrier of light, is absolutely stable and lives infinitely long; therefore, these light-years are not a hindrance for it. Light forms a light field filling space, which makes the signal propagation time finite. On its way, the light field is distorted by atmospheric turbulence, the optical systems of telescopes, and digital data processing systems. Perhaps a light ray coming to the Earth from distant stars gains something while penetrating through the cosmic vacuum, which is not empty, or through relic radiation, which carries some information about the origin of the Universe. Therefore, a human thought has to process the distorted data and information that a light ray brings to it. At the same time, apparently, these data should also be considered when forming the cognitive semantics of AI models.

Let us trace the propagation of the light field, for clarity, from the light-emitting stars. They are so far away that any of them can be considered just a point of light. Therefore, falling on the plane of the telescope objective several meters in size, the wavefront from the stars can be considered flat. This is a rather strong assumption since the wave can be swept by atmospheric turbulence and, in this case, its flat nature will be violated.

There are no perfect telescopes. Any optical system distorts the wave front. As a result, the wave beam loses its strict shape, and the rays stop converging at the focal point. The contours of the object (distant star) become blurred. This phenomenon is called aberration; it can be wavy, angular, longitudinal, and transverse. In an optical system, aberrations cannot be completely removed, just reduced. All aberrations are axisymmetric. Functional relations can be established between different types of aberrations. For this, the aberration is expanded into a series, and each term of the series has its own peculiarity and name: spherical, coma, astigmatism, field curvature, and distortion. Since light carries the range of colors, chromatic aberrations appear, the functionality of which depends on the wavelength. Lens elements are used in the telescope as correctors expanding its useful field. The refractive indices of lenses

A. Raikov, *Cognitive Semantics of Artificial Intelligence: A New Perspective*, SpringerBriefs in Computational Intelligence, https://doi.org/10.1007/978-981-33-6750-0_20

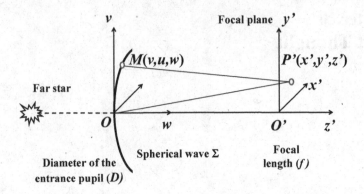

Fig. 20.1 Diagram of optical system

differ for different wavelengths. As a result, beams of different wavelengths after refraction are collected at different distances from the selected focus plane. This phenomenon is called position chromatism or longitudinal chromatism.

Each of the aberrations contributes to the distortion of the star's image. They are entirely determined by the structural elements of the optical system, such as the radii of curvature and the shapes of surfaces, the thicknesses of lens and air gaps between them, and the refractive indices of lenses (i.e., the types of glass from which they are made). When constructing telescopes, only approximate values of the structural elements of the optical system can be determined. In the aberration formulas, the expansion terms of high and even infinite order should be taken into account. Higher-order aberrations are extremely difficult to consider: they cannot be processed with absolute precision on digital (discrete) computers. The use of optical computers will be accompanied by the same distortions generated by aberrations, etc.

In the focal plane of a telescope on the side of the objects space, the image of a star undergoes the Fraunhofer diffraction. In a telescope, the image is presented to the observer as a bright disk surrounded by many rapidly decreasing rainbow rings. A bright disk carries about 85% of the energy. This aspect was described in detail long ago, in classical literature; for example, see [1].

The optical system of a telescope builds a distorted image of an almost infinite point on its main axis. Let D be the diameter of the entrance pupil of the telescope objective, f be the focal length of the telescope, λ be the wavelength of light, A be the relative aperture of the telescope objective, and V be its reciprocal. Then in the ideal optical system operates as shown in Fig. 20.1; also, see [2].

In the exit pupil, let us introduce the coordinate system (u, v, w), and in its focus, the coordinate system (x', y', z'). According to the Huygens–Fresnel principle, each point of the spherical wavefront Σ from the exit pupil of the telescope objective is itself a source of spherical waves interfering with each other. For example, from the point M of the wavefront to the point P' located in the output focal plane, a wave will come in the phase $\varphi = (2\pi/\lambda)MP'$, where the bracketed expression is the wave number. Assume that the reference is the oscillation caused at the point P' by the ray coming from the origin O. Then the disturbance at P' that has come from M will be equal to $e^{\frac{2\pi i}{\lambda}\Delta}$, where $\Delta = MP' - OP'$. Then, the oscillation caused by the spherical wave will be represented by a double integral of the disturbance over the entire spherical wavefront Σ.

In this case, the disturbance at the points of the sphere can be generally described by some *pupil function*. If the exit pupil of the telescope is round and the pupil function is equal to 1 (ideal case), then the frame of polar coordinates becomes convenient. As a result, the integral reduces to the Bessel function, which demonstrates the following fact: the intensity distribution in the image of the luminous point obeys some regularity with the amplitude of the light oscillation at the entrance pupil [3]. In the center of the image, the intensity has the greatest value (Airy disk); when displaced, it rapidly decreases, reaching 0 at some distance (dark ring). This ring is followed by alternating bright rings of decreasing intensity, which form a diffraction pattern.

The intensity of the Airy disk is proportional to the square of the entrance pupil diameter and the square of the relative telescope aperture. This relation seems obvious: the amount of light collected is proportional to the entrance pupil area of the telescope, and the image size is proportionally reduced. The radius of the first dark ring of the Airy disk as known is

$$r_0 = 0.61\lambda/n\sin\alpha, \qquad (20.1)$$

where λ is the wavelength, n is the refractive index of the medium, α is the angle subtended at the edge of the objective aperture. Since light covers different wavelengths, and the diameter of the core and rings depends on the wavelength, the ring-shaped image is rainbow-colored. In professional telescopes, during mathematical transformations, it is necessary to consider the higher-order infinitesimal terms in the expansion of the signal [4]. In particular, this is dictated by the need to take into account the dependence shown in [5]: with an increase in the relative aperture of the telescope, the radius of the first dark ring decreases, and the amount of energy in the bright rings increases. As was substantiated in [6], the defocusing tolerance should be tightened with increasing the telescope aperture.

Diffraction effects have an adverse impact on the resolving power of the telescope. Two adjacent stars will be observed in the telescope as one. Such a problem may arise when trying to find *cosmic strings* (CSs) preserved from the relic lifetime of the Universe [7]. An approach to detect such a string could be as follows: passing by a CS, the light from one distant star possibly splits; then two adjacent and absolutely identical stars will become visible in the telescope. But even if this is the case, are modern telescopes equipped with mathematical apparatus and AI tools capable of providing the necessary resolving power? After all, the light from the stars is incoherent and distorted, and the diffractive perception of the star by the telescope prevents good resolution.

Fairly strong telescopes allow one to see only the diffraction images of two stars or their spawned twins. Therefore, at best, these images will appear only partially separated or, at low resolution, merged into one star. The first signs of darkening in the diffraction pattern occur at an angular distance x between the image centers of two stars of the same brightness at $x = 0.78\varphi$, where φ is given by (20.2). With a further increase in the angle φ, the gap depth will increase, and the resolution will improve. Confident resolution occurs when the centers of the images of the stars and the gap contrast by 5%. However, according to Lord Rayleigh's conclusions, a good resolution is achieved when the maximum image of one star falls on the first dark ring of the second star. In this case, the contrast of the images of the stars and the gap reaches 26%. The angle between the stars, corresponding to this case, is equal to

$$\varphi(\text{rad}) = 1.2197\lambda/D. \qquad (20.2)$$

The reciprocal of this value is known as the Lord Rayleigh resolution of the telescope. At the same time, note that this resolution may not be enough for solving separate problems. The main reason is that 74% of the image intersection will be accompanied by turbulent and diffraction distortions, interference aliasing, etc.

When recognizing a stellar pair, it is necessary to consider the basic theorem of physical optics (for an ideal telescope), which states the following: the Fourier transform of the image of an incoherent extended object is the product of the Fourier transform of the object itself and the Fourier transform of the image of a point object. This theorem expresses the connection between the image of an extended object and the object itself, depending on the image of a point object, produced by a given optical device.

The diffraction image of stars in white light introduces its own distortions. This fact is especially important for a refractor, in which there always exists a secondary spectrum. For determining the total visible intensity of the polychromatic image of a star, one should take into account the secondary spectrum of the objective, the distribution of energy in the spectrum of the star, and the color sensitivity of the light detector.

In the case of a ground-based telescope, turbulent atmospheric phenomena leave their mark on the image quality. The refractive index of air depends on its density (pressure) and temperature; it varies along the path of a light ray in the atmosphere. This causes the ray to bend (the so-called refraction). Due to refraction, celestial bodies seem to be somewhat elevated above the horizon.

In the telescope's field of view, the stars are located at different zenith distances, and each of them has a specific displacement. This phenomenon is called differential refraction. The daily rotation of the firmament leads to a continuous change in the zenith distance of the stars and, consequently, to a change in refraction. With a high exposure time of observation (photographing), the stars located at the edge of the field of view will slightly stretch out into dashes. This restricts the shutter speed (exposure), impairing the ability to extract the useful signal from the noises.

The above features of optical transformations affect the recognition accuracy, e.g., of paired space objects. The recognition of the CSs (see the discussion above) requires comparing the spectra of the images of a pair of nearby stars. Hence, if paired stars are detected to show identical spectra and even the same redshift, one cannot exclude that this happened by chance due to the existing distortions. For a more accurate assessment of identical stars in a pair, it is necessary to reveal special cut-surfaces on the isophotes of optical images. For this purpose, images with a very high angular resolution are required: their angle must significantly exceed the threshold value (20.2).

The algorithm for detecting CSs in the Cosmic Microwave Background (CMB) maps was developed and simulated [7]. This algorithm is based on the convolution procedure of original observational radio data with the Modified Haar Functions (MHFs) and can achieve the resolution for the deficit angles ϕ of CSs of an order of ≈ 1 arcsec.

The deficit angle ϕ is an astrophysical characteristic of a CS. It appears in the lensing effect on background galaxies, making identical pairs of images on both sides of the CS, and can be given by $\phi = 8\pi G \mu / c^2$. Here μ denotes the total CS mass per unit length (linear density) that is proportional to the square root of the CS energy (for the GUT scale, the CS energy will be 10^{16} GeV); G is the Newtonian gravitational constant (6.674×10^{-11} m^3 kg^{-1} s^{-2}); finally, c is the velocity of light.

However, according to Lord Rayleigh's conclusions, an ideal telescope with a diameter of 140 mm can be used to distinguish quite clearly a binary star with a distance of 1" between its components; a telescope with a diameter of 1.4 m, a stellar pair with a distance of 0.1", provided that the stars in the pair of nearby stars have the same shine. This is about 4.8×10^{-7} rad, which is more than 7.36×10^{-7}. This double star resolution does not eliminate all distortions, which requires appropriate decoding.

Thus, the components associated with the light signal can contribute to the cognitive semantics of AI models, which are applied for solving the problems mentioned in this chapter. Sufficiently accurate decoding of the optical signal is required, which cannot be implemented by the modern methods and means available on the Earth, including both digital and optical ones. Perhaps this decoding will be developed by considering the quantum component of the optical signal. Let us dwell on this issue in detail.

References

1. Born, M., Wolf, E.: Principles of Optics. Electromagnetic Theory of Propagation, Interference and Diffraction of Light, 7th ed. Cambridge University Press (2013). https://doi.org/10.1017/CBO9781139644181
2. Maksutov, D.D.: Astronomicheskaya Optika (Astronomical Optics), 2nd ed. Leningrad, Nauka, (in Russian) (1979)
3. Airy, G.B.: On the Diffraction of an Object-Glass with Circular Aperture. Trans. Cambr. Phil. Soc. 5, 283–291 (1835)
4. Strehl, K.: Theorie Des Fernrohrs: Auf Grund Der Beugung Des Lichts (Telescope Theory: Due to Diffraction of Light). Wentworth Press (2018)

5. Hopkins, H.H.: Proc. Phys. Soc. Lond. **55**(308), 116 (1943)
6. Richards, B.: In: Kopal, Z. (ed.) Astronomical Optics and Related Subjects. North-Holland Publishing Company, Amsterdam (1956)
7. Sazhina, O.S., et al.: Optical Analysis of a CMB Cosmic String Candidate. MNRAS **485**, 1876–1885 (2019). https://doi.org/10.1093/mnras/stz527

Chapter 21
Quantization of Thought

If a thought is characterized by a wave, then it can be quantized. This is an immanent property of particles. The charges of almost all observed particles (e.g., quarks are not observed in a free state) represent multiples of the so-called "elementary charge," equal to the electron charge. To tell the truth, the discreteness of the electric charge itself is not yet clear.

Let us discuss how the phenomenon of quantization is investigated in quantum optics [1]. A quantized harmonic oscillator is associated with each mode of the radiation field. A mode is a stable state of the electromagnetic field inside a fiber or optical resonator. This is a solution of Maxwell's equations.

There may even be an oscillator corresponding to near zero energy (vacuum fluctuations). A vacuum fluctuation is a random change in the amount of energy at a point of space, as stated by the Heisenberg uncertainty principle. Some analogies with the concept of virtual particles can be useful for explaining spontaneous thought processes. This concept naturally arises because, according to the principle of particle-wave dualism, any interaction between elementary particles consists in the exchange of field quanta, e.g., the exchange of photons between an electron and a proton. But a free electron cannot emit and absorb a photon: such an event contradicts the law of conservation of energy. This paradox can have several solutions. For example, one solution is an exchange of virtual photons that transfer momentum but not energy.

However, let us revert to quantum waves. A scalar field description has been presented above; see Eq. (12.1). In a vector representation, the electric field strength **E** is described by a similar wave equation:

$$\Delta \mathbf{E} - \frac{1}{c^2} \frac{\partial^2}{\partial t^2} \mathbf{E} = 0, \tag{21.1}$$

where c denotes the velocity of light.

In a dynamic problem specified by Maxwell's equations, the momentum of the jth mode is represented in the form $p_j = m_j \dot{q}_j$, where m_j is some constant of the dimension of mass associated with a mechanical oscillator and q_j is a coordinate in the Cartesian space. (A point above any variable denotes its time derivative.) Then quantization in this problem can be implemented by assuming that p_j and q_j are operators satisfying the commutation relations:

$$[q_j, p_{j'}] = i\,\hbar\delta_{jj'}, \tag{21.2}$$

$$[q_j, q_{j'}] = [p_j, p_{j'}] = 0. \tag{21.3}$$

If the field is considered in a sufficiently large but still bounded cavity (a space with indefinite boundaries), then the matter concerns a traveling wave in this space. In the classical case, it can be characterized by the presence of transverse components with two (magnetic and electric) polarization directions, density, boundary conditions, and the series expansion in an infinite discrete set of wave vector values (taking into account the complex conjugation of the series terms). Moreover, a canonical representation of the expressions (21.2) and (21.3) becomes convenient for calculations:

$$[a_j, a_{j'}^+] = \delta_{jj'}, \tag{21.4}$$

$$[a_j, a_{j'}] = [a_j^+, a_{j'}^+] = 0, \tag{21.5}$$

where a_j and a_j^+ are called the annihilation and creation operators, respectively. An important consequence of applying these quantum conditions is that the strengths of the electric and magnetic fields cannot be measured simultaneously since the corresponding operators do not commute with each other. More specifically, their parallel components can be measured simultaneously, but the perpendicular ones cannot.

Recall that a quantum state is any possible state of a quantum system. In a closed quantum system (the solutions of the Schrödinger equation), states mean only stationary states—the eigenvectors of the Hamiltonian corresponding to different energy levels. The eigenvalues of the energy operator are discrete in contrast to the classical electromagnetic theory. In this case, the average energy can take different values: generally speaking, the state vector is an arbitrary superposition of the energy eigenstates:

$$|\psi\rangle = \sum_n c_n|n\rangle, \tag{21.6}$$

where c_n are some complex coefficients. Note that nonzero fluctuations also exist for the vacuum state $|0\rangle$, thereby stimulating the spontaneous emission of excited atoms. If the field involves many modes, then more general states can also be generated, including correlations between the field modes, which arise due to the interaction of various field modes with the quantum system as a whole.

Nonzero fluctuations and corrections caused by the interaction between an electron and vacuum will change the ratios of the energy levels of electrons in atoms. These changes can reach significant values, up to an order of 1 GHz (the so-called Lamb shift). Initial attempts to calculate the corrections led to infinite level shifts, making the solutions divergent. However, the experiments of Lamb and Rutherford generated a chain of advances in understanding quantum phenomena, the emergence of quantum electrodynamics, renormalization theory, stochastic electrodynamics, etc.

To be comprehended, quantum phenomena often require a classical description or are presented in a semiclassical form. For example, the last of the theories mentioned in the previous paragraph can be treated semiclassical. According to this theory, matter is considered quantum-mechanical, and radiation is described by classical wave equations, with vacuum fluctuations added. Such an approach was expected to describe important quantum phenomena (like spontaneous emission and the Lamb shift) and even to take a step towards the theory of everything. However, these expectations turned out to be not completely true. The phenomenon of quantum beats showed that the results of quantum computation are very different from the results obtained using the semiclassical theory, even supplemented by the concept of vacuum fluctuations. This is a good example of a complex problem that cannot be explained, even qualitatively, within the semiclassical theory, not to mention calculations.

Quantum measurement theory gives an understanding of the physics of quantum beats, their tone, amplitude, etc. For example, in three-level quantum systems, a coherently excited atom goes to a lower energy level (atomic relaxation). There can be two configurations of the atom: (a) two electrons have higher energy than the third; (b) one electron has higher energy than the other two. In case (a), the relaxing electrons emit a photon at one of two different frequencies. In case (b), one coherently excited atom relaxes, emitting a photon at one of two frequencies. However, in this case, a period of time after relaxation, observation of the atom could provide information on the relaxation path, and no beats occur. Both transitions lead to the same final atomic state, and hence it is impossible to determine on which path the atom actually relaxed. Similar to the well-known double-slit problem, this uncertainty leads to interference between photons at two frequencies, thereby causing quantum beats.

In case (b), there can be an association with the entanglement effect; see Chap. 13. Not the locations of atoms in a certain coordinate system, but quantum states play a decisive role in quantum theory. Hence, this effect determines the nonlocal nature of the cognitive semantics of AI, reflecting the fact that after measuring the quantum state of one of the entangled quantum particles, the state of another particle becomes known instantly. Moreover, the distance between these particles can be very large.

The phenomenon of quantum beats shows the limitations of the semiclassical theory supplemented by vacuum fluctuations. In other words, the statements that "photons interfere only with themselves" should be applied only to the experiments like the "double slit."

Thus, in the semiclassical theory, one system includes the wave description of (classical) radiation and (quantum) matter. With this consideration, one can quantize the radiation, and the equations for this system can be considered from a unified standpoint. Since light can be polarized, the vector nature of the field can be neglected, for the sake of simplicity and whenever necessary.

When passing from the classical description to the quantum representation, the coefficients of the eigenfunctions of the field are replaced by quantum field operators. The same quantization procedure applies to matter. For example, the wave function of a system with mass (atom, electron, etc.) is described by a superposition of states:

$$\psi(\mathbf{r}, t) = \sum_p c_p e^{-i v_p t} \phi_p(\mathbf{r}), \tag{21.7}$$

where $v_p = E/\hbar$ gives the probability amplitude, which is a number used for describing the behavior of the particle in the state $\phi_p(\mathbf{r})$.

In this case, a secondary quantization procedure is introduced, which consists in replacing each probability amplitude c_p with the annihilation operator $\widehat{c_p}$, which obeys the Fermi–Dirac commutation relations (e.g., for an electron) or the Bose–Einstein commutation relations (e.g., for a photon). Then the wave function becomes an operator that annihilates the particle at the point $\mathbf{r}$, and the state of the system is described by the vector $|\psi\rangle$.

The situation can be viewed from the other side. Let photons, mesons, bosons, etc., be represented in the environment of a quantum field. Then the wave function of the particle is obtained from the state vector by constructing the scalar product of the eigenvector of the position operator $|\mathbf{r}\rangle$ and the state vector $|\psi(t)\rangle$:

$$\psi(\mathbf{r}, t) = \langle \mathbf{r} | \psi(t) \rangle. \tag{21.8}$$

So the semiclassical theory of radiation and matter considers fields using the equations of Maxwell and Schrödinger. Both fields demonstrate a wave character, but $\hbar$ is used only when quantizing the equation for matter. A fully quantum theory, e.g., from the standpoint of the Dirac–Schwinger condition, gives a unified description for radiation and matter.

The creation operator, acting on a vacuum, due to fluctuations, can create a particle at the point $\mathbf{r}$. This allows using the conventional expression for the wave function for matter as well. At the same time, it should be emphasized that "the wave function of a photon" differs from the wave function of a nonrelativistic particle. One should be very careful with the representation of "a photon as a particle," at least when considering a one-photon phenomenon.

Thus, there is a deep relationship between quantum mechanics and the mechanics of quantized harmonic oscillators. As Dirac argued, for many identical bosons they represent exactly the same systems. An oscillator corresponds to each independent bosonic state. This makes it possible to combine the wave and corpuscular theories of light.

In a laser, the wave packet is coherent. A coherent wave packet has minimum uncertainty and is similar to a classical field. The corresponding state vector of quantum particles in such a packet is a coherent state $|\alpha\rangle$, which is the eigenstate of the positive-frequency part of the electric field operator or, equivalently, the eigenstate of the field annihilation operator. This is how a classical field behaves when waves have a definite amplitude and phase. The things are quite different in a quantum field.

A field with $|\alpha\rangle$ particles has a well-defined amplitude but a completely uncertain phase. It has the same uncertainty values for the two variables. At the same time, various effects, such as vacuum and thermal fluctuations and stochastic effects, distort the picture. Coherence becomes partial. New distributions appear, e.g., the Wigner–Weyl distribution (W-distribution), the Glauber–Sudarshan P-distribution, and the Q-distribution. They correspond to different types of the ordering of the operators a and a^+; see the discussion above.

The first of the listed distributions is intended to replace the wave function that appears from the Schrödinger equation with a symmetric probability distribution function. The P-distribution is associated with calculating the normally ordered correlation functions of the operators a and a^+. The Q-distribution allows one to find the antinormal ordering of the correlation functions. These distributions construct a bridge between quantum and classical theories of coherence. A relationship can be established between these distributions: there exists a generalized distribution of the density operator containing the three distributions as special cases.

Distribution functions allow minimizing the uncertainty of a field in a coherent state. For example, the probability that n photons are in a coherent state $|\alpha\rangle$ obeys Poisson's law. The field can be described with some degree of approximation using conjugate components. The uncertainties of the two conjugate variables characterizing the state of a particle satisfy the Heisenberg.

Thus, the construction of the cognitive semantics of AI models implies taking into account the wave-particle aspects of the field theory, which cannot be fully described within the classical and even semiclassical theory. The complexity of the wave nature inherent in cognitive semantics suggests various approaches to measure or represent such semantics using the methods and means of appropriate complexity, e.g., using interferometry tools.

Reference

1. Scully, M.O., Suhail Zubairy, M.: Quantum Optics. Quaid-i-Azam University, Islamabad (1997). https://doi.org/10.1017/CBO9780511813993

Chapter 22
Thought Interferometry

If the tools for supporting human consciousness exist and have the nature of a physical field, they cannot but respond to signals coming into their field of action, including the signals received from distant sources. If so, the signals should undergo interference, and then a natural hypothesis is that this interference can be detected.

The phenomenon of interferometry, including quantum interferometry, based on the correlation of the field amplitude and photon intensity, underlies the construction of devices for studying different objects of the micro- and macroworld, as well as the phenomena of wavy, quantum, and relativistic nature.

The quantum correlation functions of the field are substantiated by the theory of photodetection. Many observable quantities (statistics of photoelectrons, the spectral distribution of the field, etc.) are associated with the correlation functions of the field. They are necessary both to describe the quantum two-slit experiment and to fix the spectra of light from stars. Note that the correlations of intensities for different components of the studied phenomenon are measured.

Quantum coherence theory investigates field states with non-classical statistical properties. For example, the four-dimensional Minkowski space allows taking into account the curved situation for solving gravitational problems. According to the general theory of relativity, gravity is represented as a curvature of the four-dimensional space. It arises due to the presence of massive bodies in the cosmos. In earlier times, it was impossible to test this theory experimentally because of the infinitesimal value of the gravitational constant G. However, thanks to the development of laser and quantum optics, new methods and means of interferometry, astrophysical and relativistic experimental studies have become clearly more accessible. There are many interferometers, e.g., the Michelson interferometer, the Henbury–Brown–Twiss (HBT) interferometer, and ring laser resonators, to name a few.

As is well known, the Michelson interferometer determines the phenomenon of gravity by measuring the differences in the path lengths traveled by light in the arms of the interferometer and estimating the phase shift. A phase shift can occur

© The Author(s), under exclusive license to Springer Nature Singapore Pte Ltd. 2021   113
A. Raikov, *Cognitive Semantics of Artificial Intelligence: A New Perspective*,
SpringerBriefs in Computational Intelligence,
https://doi.org/10.1007/978-981-33-6750-0_22

in two cases: the gravitational wave changes the distance between the mirrors, and the gravitational wave changes or perturbs space-time, acting as a dielectric. To increase the effective optical path, resonators are used that repeatedly reflect a light beam passing through them. In this interferometer, the measurement of gravitational waves is based on the expansion or contraction of a cylinder located in the direction of propagation of the radiation. The changes in path length and phase displacement are used for making necessary estimates.

The HBT interferometer involves two photodetectors. This configuration is also possible in a hypothetical quantum-wave model of the AI's cognitive semantics. If one multiplies the detector currents coming from two photodetectors, then a low-frequency interference term appears in complex mathematical expressions. In this case, there are no individual terms sensitive to atmospheric turbulence. Under these conditions, the quantum theory of photon detection and correlation works well.

In the stationary theory of gravity, the scalar potential satisfies the Laplace equation, and in the nonstationary theory, the wave equation of motion (12.1). According to these equations, gravitational effects propagate at the velocity of light from sources in the cosmos (e.g., binary stars, including those formed by cosmic strings, exploding galaxies, etc.). The received light rays enter the telescope; after many distortions (due to atmospheric turbulence, diffraction, aberration, etc.), they finally reach a laboratory on the Earth. In the laboratory, "the remnants of truth" are digitized, calibrated, filtered, and processed on computers. As a result, a "gravitational wave" is diagnosed by studying the dynamics of oscillations with a very small amplitude and distortions (no more than 10^{-21} m).

Different theories of gravity have been elaborated, which lead to different predictions of gravitational effects. For a systematic comparison of these theories with experiments, the theoretical foundations of studying a wide class of metric theories of gravity under weak field conditions have been developed.

For example, a supersensitive ring laser resonator placed on the surface of the Earth can detect several rotations. Besides the rotation of the ring and rotation of the Earth, there are three other contributions, respectively. The first of the rotations is regarded as a "weak" confirmation of the well-known Mach principle: the inert properties of a body depend on the mass and location of other bodies. In other words, a gyroscope experiences rotation even being fixed relative to stars, and also if the Earth's motion is taken into account and does not affect the ring resonator. This rotation occurs due to the fact that the detector is located near another massive body (the Earth). It can also be viewed as a variant of magnetic gravity, analogous to the magnetic moment of a rotating electron. The second rotation occurs due to the presence of a separate coordinate system. It can be associated with the existence of the background radiation of some black body. This effect is least substantiated in the physics of gravity. Another rotation is due to the use of a curved spatial metric. This physics of space leads to the curvature of light rays from stars, gravitational redshift, etc.

The Michelson interferometer can be used for measuring the angular distance between beams from two distant stars. For this purpose, light is collected by means of two mirrors and then directed into a photodetector, which is located at the same distance (path length) to the mirrors. Light is filtered so that the frequencies of light from each star can be considered the same. Such an assumption seems justified for confirming the hypothetical existence of cosmic strings with the effect of "splitting" of light from galaxies by a cosmic string: the light spectra of compared stars should be the same [1]. With identical stars, other simplifications can be introduced in complex calculations. For example, if radiation is considered thermal, then some wave components of light can be neglected. In this case, it can be demonstrated that the photocurrent from the stars will contain an interference component, which can be investigated by changing the distance of the light paths from the two mirrors to the photodetector.

In the optical range, the operator describing the field can be decomposed into positive- and negative-frequency parts. Within this range, detectors based on the photoelectric effect are used for field measurements. An atom in the ground state is placed at some point of the radiation field, and then photoelectrons arising due to photo-ionization are observed. In such detectors, measurements are destructive because the photons responsible for the production of photoelectrons disappear. The probability of the transition of the detecting atom to an excited state can be determined, which occurs due to absorbing a photon at the given field point for a certain period of time. Note that the initial and final states of the field can be found. In practice, the former is inexactly known, being determined on a statistically stationary basis; the latter is never measured, only calculated.

This example is interesting for an indirect approach to determine the final state of an event, which is immeasurable due to the quantum features of the field. Coherences can have a different order of complexity. For example, the quantum two-slit experiment is an effect of the first order; the incoherent excitation of atoms, studied using the HBT effect for thermal and laser radiation, is an effect of the second order.

Well, there are appropriate methods and tools to confirm the hypothesis on, as well as to identify the features of, the field behavior of human consciousness and thinking, and hence to construct the cognitive semantics of AI. However, these methods and tools should be further developed and adapted towards improving their capability to detect very weak signals subjected to the collapse effect when trying to be observed.

Reference

1. Sazhina, O.S., et al.: Optical Analysis of a CMB Cosmic String Candidate. MNRAS **485**, 1876–1885 (2019). https://doi.org/10.1093/mnras/stz527

Chapter 23
Architecture of AI Semantics

Traditional AI methods and tools are based on formalized approaches, such as logical predicates, ontologies, artificial neuronal networks, ant algorithms, and others. These approaches deal with tangible things and have a discrete or digital character. They help to create denotative semantics. Simultaneously, human consciousness reflects non-formalizable cognitive phenomena: feelings, thoughts, free will, etc. These phenomena have infinite, continuous, quantum, wave, and nonlocal nature. Therefore, the cognitive semantics of AI has to be created using approaches other than the traditional ones.

For example, as is well known, the behavior of a natural neural network is random. This randomness covers profound levels of mind, including subatomic ones. An atom can emit electromagnetic radiation spontaneously. This process cannot be explained from the viewpoint of traditional quantum mechanics, where the energy levels of particles are quantized, but there is no quantization of the electromagnetic field. In the traditional approach with the Schrödinger equation, the excited states of atoms must remain stationary since this equation does not explain spontaneous emission. The cause of spontaneous emission can be the interaction of quantum particles with zero oscillations of the field in a vacuum. This phenomenon cannot be captured using the traditional semantics of AI generated by logic or even deep neural networks.

At the same time, one can try to represent the spontaneous cognitive phenomena of human consciousness in AI models using the mechanisms considered in the quantum field theory and the quantum theory of gravity. These systems will accordingly cover all kinds of fields: electromagnetic, gravitational, weak, and strong. It might seem that the AI model of a quantum system is simplified, e.g., due to neglecting the small values of gravitational force. However, the gravitational field is an attribute of AI's nonlocal semantic, and it cannot be thrown out of consideration.

The atomic- and universe-level elements can be the subject of the cognitive semantics of AI models. From such a premise, Fig. 23.1 illustrates the architecture of the cognitive and denotative semantics of AI models. The concept of cognitive semantics can be adopted to differentiate "strong AI" from "weak AI." Weak AI systems

A. Raikov, *Cognitive Semantics of Artificial Intelligence: A New Perspective*,
SpringerBriefs in Computational Intelligence,
https://doi.org/10.1007/978-981-33-6750-0_23

Fig. 23.1 Cognitive and denotative semantics of AI: architecture

only process symbols without a non-formalizable understanding of what they mean. This is a traditional look at AI systems, which can also be "narrow" and "general." A "narrow AI" system solves a particular problem better than humans would do. An artificial "general" intelligence (AGI) system can solve problems in different application domains, with better results than humans. Sometimes AGI and strong AI are considered as synonyms. Therefore, weak AI uses only denotative semantics. Strong AI processes symbols, as well, but it can understand what they mean owing to cognitive semantics. Previously, the difference between "weak AI" and "strong AI" has been most important to philosophers and practitioners irrelevant to AI. With the concept of cognitive semantics, strong AI systems can become a kind of cyber-physical systems, which have engineering character and can be used in different sectors of the economy.

The architecture presented in Fig. 23.1 is needed in theoretical and practical tools for future development. These tools can be based on the neuromorphic paradigm of AGI, which involves the method of "reverse engineering" and copying of the neural structure of the brain [1]. The neuromorphic engineering approach was suggested in the late 1980s. This concept suggests creating an electronic analog of natural neuro-biological structures. This paradigm has absorbed all the advantages of the previous paradigms of weak AI that have a discrete character. But strong AI does not focus only on the discrete and digital interpretations of the processes of thinking and consciousness; instead, strong AI places a stake on the representation of the human brain and body to the subatomic level, taking the relativistic and nonlocal effects into account.

Using quantum physics tools for representing the cognitive semantics of AI models requires increasing the dimension of the space of semantic interpretation to infinity. The aspects of quantum theory and the theory of relativity have to be considered properly. For more complete and detailed coverage of these aspects in the cognitive semantics of AI models, mathematical group theory, and category theory with the inverse problem-solving method can be used.

The elements of spatial sets will be various entities: events, ideal and virtual objects, complex numbers, and vectors with their scalar products. With each event a point in space can be associated, treated as a mathematical object. The Fourier transforms over the elements of the multidimensional event space can be applied, facilitating the use of optical mechanisms to interpret the cognitive semantics of AI.

With this approach, the AI symbolic model, including its formalized denotative semantics and non-formalized cognitive semantics, can be placed in some single space, and then unified operators can be used to implement the transformations of various-nature events. If cognitive semantics allows for a quantum interpretation, with its representation in the Hilbert space, and a relativistic interpretation, with its representation in terms of the transformation of the Calabi–Yau manifolds [2], then group theory can be fruitful for solving different issues of nonlocal cognitive semantics.

The cognitive approach to constructing the semantics of AI models is already of practical use in the real sector of the economy and social life.

References

1. Fjelland, R.: Why General Artificial Intelligence will not be Realized. Humanit. Soc. Sci. Commun. **7**, 10 (2020). https://doi.org/10.1057/s41599-020-0494-4
2. Yau, S.-T., Nadis, S.: The Shape of Inner Space: String Theory and the Geometry of the Universe's Hidden Dimensions. Basic Books. A Member of the Perseus Books Group (2010)

Chapter 24
Applications

The construction of cognitive semantics for AI models has confirmed its effectiveness in real practice: the design and application of decision support systems, the elaboration of development strategies for regional sectors of the economy. The use of cognitive semantics in real practice has improved the complete representation of discourse when solving management problems in business and government organizations. For example, when conducting strategic meetings, an appropriate procedure to structure information considerably accelerates the processes of discussing problems and reaching a collective agreement among the participants on goals and courses of action [1].

The solution of an inverse problem on a cognitive model significantly accelerates the generation of group ideas and decision-making; see [2, 3]. As a result, a draft strategy for the development of an industrial enterprise, a large corporation, a department of a ministry, etc., is elaborated via a parallel series of brainstorming sessions and strategic meetings. In the course of a purposeful discussion of strategic issues, with the participation of the company's top managers and invited experts, a vision of the company's development strategy is formed in a few hours. During this time, a tree of goals is consistently formed, a strategic analysis of the socio-economic situation is carried out, priority areas of the group's actions are identified, and a plan of strategic measures is developed.

For example, using the proposed approach, 35 brainstorming sessions were conducted in parallel in 4 h, and a cognitive model was constructed to substantiate a breakthrough strategy for the development of tourism in a megapolis [4]. It was required to change the tourism situation dramatically. A SWOT (strengths, weaknesses, opportunities, threats) analysis revealed 70 factors for representing the dynamic situation. These factors were convoluted into 15 ones, most significant; the weights of their mutual influence were evaluated; finally, the cognitive model was constructed. A Big Data analysis was employed to verify the model. The genetic algorithm was applied to answer the question: "What has to be done to achieve the strategic goal?".

A. Raikov, *Cognitive Semantics of Artificial Intelligence: A New Perspective*, SpringerBriefs in Computational Intelligence, https://doi.org/10.1007/978-981-33-6750-0_24

Manufacturing strategic planning is an ill-defined decision-making process characterized by high strategic risks; see [5]. Many factors influence the determination of the manufacturer's goals. Many aspects have to be taken into account: market requirements, the interests and desires of participants, etc. The number of factors may reach more than a hundred. The convergent approach, including cognitive modeling and the genetic algorithm, was used to accelerate assessment and reduce strategic risks. Divided into 3 blocks, 24 parameters have to be evaluated. The approach was tested in real practice for the strategic risk temperature assessment of an industrial enterprise that could not permit third-party consultants on its territory. The strategic risk assessment of an industrial enterprise can be enforced by top management for 1–2 h.

The convergent approach demonstrated its fruitfulness for developing a methodology to assess the effectiveness of the organizational structure of a national bank, whose strategic goals could not be used as assessment criteria [6]. Quantum operators were used for finding an optimal distribution of employees among different levels of the hierarchical management structure of the bank.

Quantum operators were used for improving the quality of forecasting of long-term needs for educational services under a high uncertainty about the development of the national economy [7]. When applied in forecasting, quantum operators make it possible to consider a wide range of possible scenarios, which go beyond the standard extrapolation of retrospective event dynamics.

The problems associated with the need to clean up big data are well known. They naturally arise in data mining, deep learning, forecasting, and other technologies. Big data contains too much information garbage. For example, in the book [8], the "dark" sides of data were identified. In such conditions, the proposed approach with cognitive modeling [9] accelerates the cleaning of big data, helping to select relevant data and to structure arrays in an adequate way, depending on the problems being solved.

Online meetings, organized with an increasing frequency, are characterized by an indirect contact of participants. This slows down the processes of reaching a collective agreement among the participants and hence their coordinated decision-making. The proposed tools with the cooperative construction of a cognitive model help to accelerate the networked processes of interests coordination. This is achieved by the correct structuring and convolution of information generated at meetings.

However, some practical problems remain unsolved. For example, a networked strategy meeting with cognitive modeling can be carried out only in several steps. Therefore, the following issues seem debatable: it is possible to hold a corporate strategic meeting, including the construction of a cognitive model and its verification based on the analysis of Big Data, in 2–3 h; or even do it automatically and instantly, taking advantage of cognitive semantics. This is very important in emergencies [10]. To answer this question, the potential of quantum, relativistic, and other aspects of cognitive semantics discussed in this book can be investigated.

References

1. Raikov, A.N.: Convergent Cognitype for Speeding-up the Strategic Conversation. In: IFAC Proceedings Volumes, vol. 41(2), pp. 8103–8108. Seoul, South Korea (2008). https://doi.org/10.3182/20080706-5-KR-1001.01368

2. Raikov, A.N., Panfilov, S.A.: Convergent Decision Support System with Genetic Algorithms and Cognitive Simulation. In: Proceedings of the IFAC Conference on Manufacturing Modelling, Management, and Control, Saint Petersburg, Russia, pp. 1142–1147 (2013). https://doi.org/10.3182/20130619-3-RU-3018.00404

3. Raikov, A.N., Ermakov, A.N., Merkulov, A.A.: Assessments of the Economic Sectors Needs in Digital Technologies. Lobachevskii J. Math. **40**(11), 1837–1847 (2019). https://doi.org/10.1134/S1995080219110246

4. Raikov, A.: Megapolis Tourism Development Strategic Planning with Cognitive Modelling Support. In: Yang, X.S., Sherratt, S., Dey, N., Joshi, A. (eds.) 4th International Congress on Information and Communication Technology. Advances in Intelligent Systems and Computing, vol. 1041. Springer, Singapore (2020). https://doi.org/10.1007/978-981-15-0637-6_12

5. Raikov, A.: Manufacturer's Strategic Risk Temperature Assessment with Convergent Approach, Cognitive Modelling and Blockchain Technology. IFAC-PapersOnLine **52**(13), 1289–1294 (2019). https://doi.org/10.1016/j.ifacol.2019.11.376

6. Raikov, A.N.: Organizational Structure Optimization with the Questions-Criteria Hierarchy. IFAC-PapersOnLine **49**(12), 1532–1537 (2016). https://doi.org/10.1016/j.ifacol.2016.07.797

7. Raikov, A.: Strategic Analysis of the Long-Term Future Needs of Educational Services. In: Proceeding of the 3rd World Conference on Smart Trends in Systems, Security and Sustainability, Roding Building, London, UK. IEEE (2019). https://doi.org/10.1109/WorldS4.2019.8903983

8. Hand, D.J.: Dark Data: Why What You Don't Know Matters. Princeton University Press, Princeton and Oxford (2020)

9. Raikov, A.N., Avdeeva, Z., Ermakov, A.: Big Data Refining on the Base of Cognitive Modeling. IFAC-PapersOnLine **49**(32), 147–152 (2016). https://doi.org/10.1016/j.ifacol.2016.12.205

10. Raikov A.N.: Accelerating Decision-Making in Transport Emergency with Artificial Intelligence. Adv. Sci. Technol. Eng. Syst. J. **5**(6), 520–530 (2020). https://doi.org/10.25046/aj050662

Conclusion

The rapid expansion of new AI applications has promoted a significant development in different fields of human activity. Important technological advances can be seen in integrating such novel approaches to AI-based problem-solving deep learning, big data analysis, ontologies, large-volume genetic algorithms, cognitive architectures, etc. The prediction power of AI models has been dramatically improved as the result of training on data from different sources, including Internet-of-Things systems, astronomical telescopes, and the LHC.

However, the main paradigm of AI development, resting on a logically formalized and digital representation of the semantics of AI models, remains unchanged—discrete. Even deep neural networks are basically discrete and can be represented by algorithms. This significantly restricts the ability of AI tools to represent the natural cognitive capabilities of humans, such as free will, thinking, feelings, intuition, etc. Cognitive capabilities admit neither discrete nor digital description.

New approaches and tools are needed to construct the cognitive semantics of AI models. These semantics are formed by including (immersing) a human directly in the process of solving complex interdisciplinary problems, as well as by the non-discrete emulation of human thoughts and feelings. To interpret the latter, this book has substantiated the possibility of using resonant and wave processes, optical calculations, as well as the possibility of considering the nonlocal effects of quantum field theory and the theory of relativity.

As it seems to the author, the idea of the cognitive semantics of AI opens intriguing opportunities for further advancement in the study of such problems as unified field theory, quantum gravity, dark energy and dark matter, cosmic strings and quantum superstrings, etc. This confidence is based on the new interdisciplinary approach to the development of AI tools and methods put forward in the book. This approach embraces more than a dozen different disciplines and ensures the convergence of interdisciplinary research for obtaining the necessary synergy. The solutions of complex scientific problems are searched along the purposeful and sustainable path of a comprehensive study of the properties of matter and consciousness, the quantum world, and the Universe.

This approach to the development of AI towards increasing the power of cognitive semantics that cannot be represented in a conventional digital way has already proved its effectiveness in real practice. It can be used to accelerate significantly strategic meetings in companies and government organizations, discussions of problems, and reaching a collective agreement among the participants on goals and courses of action.

A new mission of AI may sound like solving the complex inter-disciplinary problems that humankind cannot overcome with the Earth's means. For example, scientists hopelessly try to create the GUT that embraces micro- and macro-world spaces. Like the case with creating well known Dirac's equation, many such cases have to be accumulated and analyzed by the convergent way. Note that every such case may be not only an image of an object or text. It can be a complex theory with a system of proving. In particular, Dirac's equation considers the quantum and relativistic aspects of the matter, and the negative energy of fermions is paradoxically allowed by this equation. Initially, scientists thought this equation was erroneous because it seemed to contradict the fundamental laws of physics, indicating the possibility of creating a perpetual motion machine. But as it turned out, this paradox helped to explain the magnetic properties of electrons, discovering antimatter, and supposing that the cosmic vacuum is filled with negative-energy fermions. Scientists started elaborating quantum field theory, taking into account the difference between a particle and a field.

However, society, science, and industry have been setting new challenges. Modern AI, with its cognitive semantics, must assist in understanding the phenomenon of consciousness, in which, e.g., quantum fermions are filling the space of the Universe without permission to occupy the same state, while friendly quantum bosons can generate original thoughts and ideas, due to their possibility to be together in the same state. The latter phenomena are influenced by the nonlocal effects, i.e., depend on the behavior of particles of long-distant cosmic objects.

Thus, thanks to cognitive semantics, AI should not so much recognize objects but solve very complex multidisciplinary problems; not so much identify associations in big data but find the reasons of events; not only describe the external world with logic but be immersed in it.

The concept of cognitive semantics can be an approach for transforming traditional weak AI into strong AI. The former only processes symbols without an understanding of what they mean. Strong AI has to understand what the symbols mean, with the power of human-like consciousness or even more. With the concept of cognitive semantics, strong AI can become a kind of cyber-physical and engineering system, which can be used in different sectors of the economy and social life.

Glossary

Architecture of AI semantics : a hypothetical structure of artificial intelligence models' semantics that embraces its components, relationships, and the principles of its design and evolution.

Artificial general intelligence (AGI): a system capable of solving several intellectual tasks of different types simultaneously, possibly much better than humans would do.

Cognitive modeling: modeling of an ill-defined situation with identifying factors and directed interrelations between them, and taking into account their cognitive and denotative semantics.

Cognitive semantics: a semantic interpretation of artificial intelligence models by mapping them into a different kind of phenomena and spaces, taking into account the nonlocal, non-formalized, and noncausal aspects of a modeled situation.

Convergence: the process of decreasing differences, in a purposeful and sustainable way.

Convergent management: ensuring necessary conditions for the purposeful and sustainable evolution of a controlled object towards inexact goals.

Denotative semantics: a formalized semantic interpretation of artificial intelligence models.

Divergence: the process of increasing differences.

Hybrid artificial intelligence: the human-machine artificial intelligence.

Models' semantics: a semantic interpretation of the model's elements by mapping them into other objects, phenomena, and spaces.

Narrow artificial intelligence: a system capable of solving a particular intellectual task, possibly much better than humans would do.

Networked expertise (E-expertise): the process of collective expert assessment of events and objects in a distributed environment.

Nonlocal semantics: a semantic interpretation that takes into account the phenomena of the relativistic, quantum, and field nature.

Quantum semantics: a representation of cognitive semantics using special quantum operators.

© The Author(s), under exclusive license to Springer Nature Singapore Pte Ltd. 2021   127
A. Raikov, *Cognitive Semantics of Artificial Intelligence: A New Perspective*,
SpringerBriefs in Computational Intelligence,
https://doi.org/10.1007/978-981-33-6750-0

Strong artificial intelligence: a system capable of simulating human feelings, intentions, and thinking by processing symbols, fields, and other kinds of matter and energy, in order to solve complex multidisciplinary problems, with a deep understanding of what it does.

Weak artificial intelligence: a system capable of solving intellectual tasks by processing only symbols, without understanding of what it does, but faster than humans would do.

Printed in the United States
By Bookmasters